INTERVAL ORDERS
AND INTERVAL GRAPHS

WILEY-INTERSCIENCE SERIES IN DISCRETE MATHEMATICS

ADVISORY EDITORS

Ronald L. Graham
AT&T Bell Laboratories, Murray Hill, New Jersey

Jan Karel Lenstra
Mathematisch Centrum, Amsterdam, The Netherlands

Graham, Rothschild, and Spencer
RAMSEY THEORY

Tucker
APPLIED COMBINATORICS

Pless
INTRODUCTION TO THE THEORY OF ERROR-CORRECTING CODES

Nemirovsky and Yudin
PROBLEM COMPLEXITY AND METHOD EFFICIENCY IN OPTIMIZATION
(Translated by E.R. Dawson)

Goulden and Jackson
COMBINATORIAL ENUMERATION

Gondran and Minoux
GRAPHS AND ALGORITHMS
(Translated by S. Vajda)

Fishburn
INTERVAL ORDERS AND INTERVAL GRAPHS: A STUDY OF PARTIALLY ORDERED SETS

Tomescu
PROBLEMS IN CONBINATORICS AND GRAPH THEORY
(Translated by Robert A. Melter)

Palmer
GRAPHICAL EVOLUTION: AN INTRODUCTION TO THE THEORY OF RANDOM GRAPHS

INTERVAL ORDERS AND INTERVAL GRAPHS

A Study of Partially Ordered Sets

PETER C. FISHBURN

AT & T Bell Laboratories
Murray Hill, New Jersey

A Wiley-Interscience Publication
JOHN WILEY & SONS

New York · Chichester · Brisbane · Toronto · Singapore

Copyright © 1985 AT & T Bell Laboratories

Published by John Wiley & Sons, Inc.

All rights reserved. Published simultaneously in Canada.

Reproduction or translation of any part of this work beyond that permitted by Section 107 or 108 of the 1976 United States Copyright Act without the permission of the copyright owner is unlawful. Requests for permission or further information should be addressed to the Permissions Department, John Wiley & Sons, Inc.

Library of Congress Cataloging in Publication Data:

Fishburn, Peter C.
  Interval orders and interval graphs.

  (Wiley-Interscience series in discrete mathematics)
"A Wiley-Interscience publication."
Bibliography: p.
Includes index.
  1. Partially ordered sets. I. Title. II. Series.
QA171.485.F57 1984   511.3'2   84-17258
ISBN 0-471-81284-6

Printed in the United States of America

10 9 8 7 6 5 4 3 2 1

To the memory of Pearl Clingerman.

Preface

Interval orders and interval graphs have emerged during the past few decades as important research subjects in the theory of partially ordered sets and graph theory. The name *interval order* was first used in the late 1960s (Fishburn, 1970a), but the same concept was discussed much earlier as "relations of complete sequence" by Norbert Wiener (1914). The more recent name reflects the fact that a partially ordered set $(X, \prec)$ is an interval order precisely when its points $x, y, \ldots$ can be mapped into intervals in a linearly ordered set, such as $(\mathbb{R}, <)$, such that, for all x and y in X, $x \prec y$ if and only if the interval assigned to x completely precedes the interval assigned to y.

The companion term *interval graph* appeared earlier (Gilmore and Hoffman, 1962), having been previously discussed without this designation by Hajös (1957) and Benzer (1959). It refers to a graph $(X, \sim)$ whose points can be mapped into intervals in a linearly ordered set such that, for all distinct x and y, $x \sim y$ if and only if the intervals assigned to x and y have a nonempty intersection.

The close relationship between interval orders and interval graphs is suggested by two observations. First, if $(X, \prec)$ is an interval order, and if $\sim$ is the symmetric complement of $\prec$, then $(X, \sim)$ is an interval graph. Second, if $(X, \sim)$ is an interval graph, and if $\prec$ is defined from an interval representation of $(X, \sim)$ by $x \prec y$ if the interval for x completely precedes the interval for y, then $(X, \prec)$ is an interval order.

The present book is an outgrowth of my research on ordered sets. It attempts to provide a unified treatment of interval orders, interval graphs, and related concepts such as semiorders, comparability graphs, and indifference graphs (named, respectively, by R. D. Luce, A. J. Hoffman, and F. S. Roberts). The presentation is self-contained and is designed to be readily accessible to mathematicians, upper-level students of mathematics, and people in economics, statistics, psychology, computer science, and other fields who are interested in ordered sets and graphs.

My organization of the book around the theme of interval orders reflects its author's particular orientation, as does the choice of subject matter. However, important topics developed by others are included for broader and more balanced coverage. At the same time, the book does not try to treat exhaustively all of the topics mentioned, and some extensions and generalizations of interval orders and interval graphs are not discussed. A great deal of additional material will be found, for example, in the graph-theory books by Berge (1973) and Golumbic (1980), and in the recent treatise on linear orders by Rosenstein (1982).

A few remarks on personal style are in order. Over the years I have come to prefer to work with irreflexive as opposed to reflexive orders and will follow this preference here. Thus partial, interval, linear, and other types of orders symbolized by $\prec$, $<_1$, and so forth, are always irreflexive (it is never true that $x \prec x$) and asymmetric [if $x \prec y$, then not $(y \prec x)$]. Reflexive orders, when used, carry an undersymbol, as in $\precsim$ and $\leq$. The symbol $\sim$, by itself or with scripts, always denotes a symmetric binary relation; it may or may not be reflexive. In most cases, $(X, \sim)$ is a reflexive graph, so every point has a "loop" $(x \sim x)$; the main exception involves comparability graphs.

Theorems within each chapter are numbered consecutively without chapter prefix, but interchapter references add the chapter number (Theorem 5.3 is Theorem 3 in Chapter 5). The end of each proof is marked by $\Box$, $|X|$ is the cardinality of set X, $A \setminus B$ is the set of points in A but not B, $\subset$ denotes proper inclusion, $\mathbb{R}$ is the set of real numbers, $\lfloor x \rfloor$ is the integer part of x, and $\lceil x \rceil$ is the smallest integer not less than x.

For the record, and to avoid unnecessary references to my own work throughout the text, I note here publications used as source material for parts of the book. Fxy signifies Fishburn (19xy) in the references at the end of the book; numbers in parentheses are the numbers of theorems in the book that first appeared in Fxy. F69 (7.3); F70a (1.4, 2.2, 2.3, 2.8, 3.1); F70b (1.1, 1.2, 1.4, 2.2, 2.3, 2.8); F70c (1.1(c)); F71 (3.1, 4.1, 4.3, 4.5, 4.6); F73a (7.5 through 7.12); F73b (1.2(f), 2.10); F81a (9.1, 9.3); F81b (6.1 through 6.4); F82 (6.2, 6.5, 6.6); F83a (8.2, 8.4, 8.5); F83b (5.18, 7.2); F83c (10.1 through 10.6); F84a (3.12); F84b (8.1, 8.3, 9.2, 9.4); F84c (9.6, 9.7, 9.8).

I am indebted to many people for guidance and encouragement over the years in my work on ordered sets. Special thanks go to Fred Roberts, Duncan Luce, Peter Hammer, Ronald Graham, and Thomas Trotter. I am also indebted to the institutions that have supported my research in this area—The Research Analysis Corporation, The Institute for Advanced Study, The Pennsylvania State University, AT&T Bell Laboratories—and to Bell Labs for making the book possible. The entire manuscript was superbly typed by Marie Wenslau, and I thank her for invaluable assistance.

PETER C. FISHBURN

Murray Hill, New Jersey
January 1985

Contents

1 INTRODUCTION — 1

    1.1 Binary Relations — 1
    1.2 Related Sets — 2
    1.3 Ordered Sets — 3
    1.4 Linear Extensions — 7
    1.5 Transitive Reorientations — 8
    1.6 Intervals and Antichains — 11
    1.7 Graphs — 13
    1.8 Preview — 16

2 INTERVAL ORDERS — 18

    2.1 Definitions and Examples — 18
    2.2 Basic Structures — 21
    2.3 Magnitudes and Characteristic Matrices — 25
    2.4 Representation Theorems — 28
    2.5 Weak Order Extensions — 30

3 INTERVAL GRAPHS — 35

    3.1 Definitions and Examples — 35
    3.2 Basic Ties to Interval Orders — 37
    3.3 Linearly Ordered Maximal Cliques — 40
    3.4 Characterizations of Interval Graphs — 46
    3.5 Characterizations of Indifference Graphs — 50
    3.6 Unique Agreement — 52

4 BETWEENNESS — 57

    4.1 Betweenness Relations — 57
    4.2 Axioms and Preliminary Lemmas — 59
    4.3 Interval Orders and Graphs — 62

4.4	Semiorders and Indifference Graphs	70
4.5	Weak and Linear Orders	74

5 DIMENSIONALITY AND OTHER PARAMETERS   76

5.1	Parameters of Posets	76
5.2	Dimension of Posets	79
5.3	Posets of Dimension 2	85
5.4	Dimensions of Semiorders	88
5.5	Dimensions of Interval Orders	92
5.6	Breadth and Depth	95
5.7	Numbers of Related Sets	97

6 EMBEDDED SEMIORDERS AND INDIFFERENCE GRAPHS   101

6.1	Extremization Problems	101
6.2	Upper Bound and Exact Values	104
6.3	General Bounds	110
6.4	Patterns with Restricted Antichains	116

7 REAL REPRESENTATIONS   122

7.1	The Scott-Suppes Theorem	122
7.2	Linear Solution Theory	123
7.3	Probability Intervals	130
7.4	Cantor's Theorem	132
7.5	Closed-Interval Representations	134
7.6	General Interval Representations	139

8 BOUNDED INTERVAL ORDERS   144

8.1	Bounded Interval Lengths	144
8.2	Unitary Classes	147
8.3	Length-Bounded Interval Orders	148
8.4	Linear Inequalities and Forbidden Picycles	153
8.5	Reducibility Lemmas	157
8.6	Banishing Forbidden Picycles	164

9 NUMBERS OF LENGTHS   170

9.1	Finite-Lengths Classes	170
9.2	Two Lengths	173

	9.3	Many Lengths	175
	9.4	Depth-2 Interval Orders	176
	9.5	Discontinuities and Admissible Lengths	179

10 EXTREMIZATION PROBLEMS — 186

10.1	Definitions and Conjectures	186
10.2	Basic Theorems	189
10.3	Exact Values	193
10.4	Departures from Linearity	199
10.5	More Conjectures	203

REFERENCES — 205

INDEX — 211

INTERVAL ORDERS
AND INTERVAL GRAPHS

1

Introduction

This chapter has two purposes. The first is to introduce basic terminology and facts about relations, orders, and graphs that are used throughout the book. The second is to outline topics discussed later. Our study of interval orders proper begins in the next chapter.

1.1 BINARY RELATIONS

An *n-ary relation* R on a set X is a subset of X^n. Although ternary ($n = 3$) and quaternary ($n = 4$) relations will be used later, most of the relations we shall deal with are binary relations, with $R \subseteq X \times X$. Notationally, xRy means the same thing as $(x, y) \in R$, and not(xRy) or $x\cancel{R}y$ signifies $(x, y) \notin R$.

We shall say that a binary relation R on X is

reflexive if xRx for every x in X.
irreflexive if not(xRx) for every x in X.
symmetric if $xRy \Rightarrow yRx$ for all x and y in X.
asymmetric if $xRy \Rightarrow$ not(yRx) for all x and y in X.
transitive if (xRy, yRz) $\Rightarrow xRz$ for all x, y, and z in X.
negatively transitive if $xRy \Rightarrow$ (xRz or zRy) for all x, y, and z in X.
complete if $x \neq y \Rightarrow$ (xRy or yRx) for all x and y in X.

Other properties will be introduced as they are needed.

An *equivalence relation* on X is a reflexive, symmetric, and transitive binary relation on X. Given an equivalence relation E on X, X/E denotes the *set of equivalence classes* determined by E. Each class in X/E is a subset of X of the form $\{y: yEx\}$, and X/E is a partition of X. Conversely, a partition of X into nonempty subsets determines an equivalence relation E by defining xEy if x and y are in the same element of the partition.

The *composition* RS of binary relations R and S on X is defined by

$$RS = \{(x, y): xRz \text{ and } zSy \text{ for some } z \text{ in } X\}.$$

When $S = R$, we write RS as R^2. For $n > 2$, $R^n = R(R^{n-1})$. Transitivity says that $R^2 \subseteq R$.

For any binary relation R on X, let $a(R)$, $s(R)$, $c(R)$, $d(R)$, and $t(R)$ be respectively the *asymmetric part* of R, the *symmetric part* of R, the *complement* of R, the *dual* of R, and the *transitive closure* of R:

$a(R) = \{(x, y): xRy \text{ and not}(yRx)\}$.
$s(R) = \{(x, y): xRy \text{ and } yRx\}$.
$c(R) = \{(x, y): \text{not}(xRy)\}$.
$d(R) = \{(x, y): (y, x) \in R\}$.
$t(R) = R \cup R^2 \cup R^3 \ldots$.

These basic operations combine to form compound operations, such as $cd(R)$, the complement of the dual of R, $sc(R)$, the *symmetric complement* of R, and $ta(R)$, the transitive closure of the asymmetric part of R. Fishburn (1978) proves that, in addition to the empty relation $\emptyset$ and the universal relation $X \times X$, at most 110 different relations can be generated from a given relation by sequential applications of the five basic operations, and that 110 is the least upper bound. Examples of duplications for different sequences are $cd(R) = dc(R)$, $ac(R) = ad(R)$, $atcat(R) = atct(R)$, and $stct(R) = tsct(R)$.

It is easily seen that $a(R) = R \cap cd(R)$ and $s(R) = R \cap d(R)$. In addition, we have the following expressions for properties defined earlier:

symmetry: $d(R) = R$
asymmetry: $R \cap d(R) = \emptyset$
negative transitivity: $c(R)^2 \subseteq c(R)$.

When A and B are nonempty subsets of X, and R is a binary relation on X, we shall write

$$ARB \quad \text{if } aRb \quad \text{for all } (a, b) \in A \times B.$$

Similarly, when $a, b \in X$, aRB means $\{a\}RB$, and ARb means $AR\{b\}$.

1.2 RELATED SETS

The simple relational system (X, R) in which R is a binary relation on X will be referred to as a *related set*. More specifically, when E is an equivalence relation on X, (X, E) is an equivalence set; when $\prec$ is a partial order on X, $(X, \prec)$ is a partially ordered set; and so forth.

We shall say that (X, R) is reflexive (irreflexive,..., complete) if R is reflexive (irreflexive,..., complete). In addition, the operations on binary relations defined in the preceding section apply to related sets by the definitions

$$\gamma(X, R) = (X, \gamma(R)) \qquad \gamma \in \{a, s, c, d, t\}.$$

Thus the asymmetric part of (X, R) is $(X, a(R))$, the complement of (X, R) is $(X, c(R))$, and so forth.

Two related sets (X, R) and (Y, S) are *isomorphic* if there is a one-to-one mapping f from X onto Y such that, for all $x, y \in X$,

$$xRy \Leftrightarrow f(x)Sf(y).$$

Isomorphism between (X, R) and (Y, S) is expressed by $(X, R) \cong (Y, S)$. It is easily seen that isomorphism is an equivalence relation on the class of related sets.

The *restriction* of a binary relation R on X to a subset Y of X is $R \cap (Y \times Y)$, and the restriction of the related set (X, R) to $Y \subseteq X$ is $(Y, R \cap (Y \times Y))$. For convenience, we shall sometimes abbreviate $(Y, R \cap (Y \times Y))$ as (Y, R), or just Y. A subset $Z \subseteq X$ in the context of (X, R) is said to be isomorphic to (Y, S) if (Z, R) is isomorphic to (Y, S).

1.3 ORDERED SETS

A binary relation R on X is a:

partial order if R is irreflexive and transitive.
weak order if R is asymmetric and negatively transitive.
linear order if R is a complete weak order.

We refer to (X, R) respectively as a *poset* (partially ordered set), a weakly ordered set, and a linearly ordered set. I leave it to the reader to show that a partial order is asymmetric, a weak order is transitive, and a complete partial order is a linear order.

Linearly ordered sets will also be called *chains*. A *linearly ordered subset* or chain in a related set (X, R) is a $Y \subseteq X$ such that $(Y, R \cap (Y \times Y))$ is a linearly ordered set. A chain in (X, R) is *maximal* if it is not properly included in another chain in (X, R), and it is *maximum* if no chain has greater cardinality.

The distinctions among partial, weak, and linear orders are illustrated by the Hasse diagrams in Fig. 1.1. In such a diagram, either xRy if x is above y and there is a downward path from x to y, or (exclusionary) xRy if x is below y and there is an upward path from x to y. The latter orientation will be used when we write R as $\prec$.

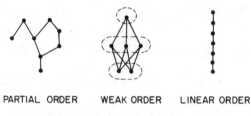

PARTIAL ORDER   WEAK ORDER   LINEAR ORDER

Figure 1.1 Posets.

When $(X, \prec)$ is a poset, we shall often use $\sim$ to denote $sc(\prec)$, in which case

$$x \sim y \quad \text{if not}(x \prec y) \text{ and not}(y \prec x).$$

We now present two theorems that record useful facts about partial orders and foreshadow some of the results proved later for interval orders. These theorems describe qualitative properties of $\prec$ with the use of $\sim = sc(\prec)$ and two other binary relations on X that are based on $\prec$. A numerical representation theorem is proved in the next section.

The first new relation is defined succinctly by

$$\approx \: = c[(\sim)c(\sim) \cup c(\sim)(\sim)],$$

where $(\sim)c(\sim)$ is the composition of $\sim$ and $c(\sim)$, and $c(\sim)(\sim)$ is the composition of $c(\sim)$ and $\sim$. In more detail, $x \approx y$ if it is false that there is a z for which $x \sim z$ and not$(z \sim y)$, and it is false that there is a z for which not$(x \sim z)$ and $z \sim y$, that is,

$$x \approx y \quad \text{if } \{z: z \sim x\} = \{z: z \sim y\}.$$

When $(X, \prec)$ is a poset, Theorem 1 notes that $x \approx y$ if x and y have the same upper sets $(\{z: x \prec z\} = \{z: y \prec z\})$ and the same lower sets $(\{z: z \prec x\} = \{z: z \prec y\})$.

The second new relation, called the *sequel* of $\prec$, is defined by

$$S_0(\prec) = [(\sim)(\prec) \cup (\prec)(\sim)] \cap cd\,[(\sim)(\prec) \cup (\prec)(\sim)],$$

where $(\ )(\ )$ denotes composition. Thus, $xS_0(\prec)y$ if either $x(\sim)(\prec)y$ or $x(\prec)(\sim)y$, and neither $y(\sim)(\prec)x$ nor $y(\prec)(\sim)x$. When $(X, \prec)$ is a poset, so is $(X, S_0(\prec))$; moreover, $\prec$ is included in $S_0(\prec)$.

Here and later X is assumed to be nonempty. Unless it is noted otherwise, there is no restriction on the cardinality of X apart from $|X| > 0$.

Theorem 1. *Suppose $(X, \prec)$ is a poset. Let $\sim = sc(\prec)$, with $\approx$ and $S_0(\prec)$ as defined above. Then*

(a) *$\approx$ is an equivalence relation.*
(b) *$\prec$, $(\prec)(\approx)$, and $(\approx)(\prec)$ are identical.*
(c) *$x \approx y \Leftrightarrow [\{z: x \prec z\} = \{z: y \prec z\}, \{z: z \prec x\} = \{z: z \prec y\}]$.*
(d) *$(X/\approx, \prec)$ is a poset.*
(e) *$S_0(\prec)$ is a partial order that includes $\prec$.*

Proof. (a) Since $\prec$ is irreflexive, its symmetric complement $\sim$ is reflexive and symmetric, and $\approx$ is clearly reflexive and symmetric. Suppose $x \approx y$ and $y \approx z$. Then $\{w: w \sim x\} = \{w: w \sim y\} = \{w: w \sim z\}$, hence $\{w: w \sim x\} = \{w: w \sim z\}$, hence $x \approx z$, so $\approx$ is transitive.

Proofs of (b), (c), and (d) are left to the reader.

(e) Suppose $x \prec y$. Then $x \sim x \prec y$, and either $y \sim z \prec x$ or $y \prec z \sim x$ yields a contradiction by transitivity, so $xS_0(\prec)y$. By its definition, $S_0(\prec)$ is irreflexive. To verify that it is transitive, suppose $xS_0(\prec)y$ and $yS_0(\prec)z$. Contrary to transitivity, suppose not($xS_0(\prec)z$). Then either

(i) $z \sim a$ and $a \prec x$ for some $a \in X$, or
(ii) $z \prec a$ and $a \sim x$ for some $a \in X$, or
(iii) there is no $b \in A$ such that either $(x \sim b, b \prec z)$ or $(x \prec b, b \sim z)$.

If (i) holds, then $a \prec y$ by $xS_0(\prec)y$, but $a \prec y$ and $z \sim a$ contradict $yS_0(\prec)z$. A similar contradiction follows from (ii). Suppose then that (iii) holds. Assume first that $xS_0(\prec)y$ is realized in part by $(x \sim c, c \prec y)$. Then (iii) implies not($c \prec z$), so either $z \prec c$ or $z \sim c$. However, since $c \prec y$, each of $z \prec c$ and $z \sim c$ contradicts $yS_0(\prec)z$. A similar contradiction obtains if $(x \prec c, c \sim y)$, so (iii) leads to a contradiction in any event, and we conclude that $xS_0(\prec)z$. □

Theorem 2. *Suppose the hypotheses and definitions of Theorem 1 hold. Then the following are mutually equivalent:*

(a) *$(X, \prec)$ is a weakly ordered set.*
(b) *$\sim$ is transitive.*
(c) *$\sim = \approx$.*
(d) *$\prec = (\prec)(\sim) = (\sim)(\prec)$.*
(e) *$(X/\approx, \prec)$ is a linearly ordered set.*

Moreover, if X is finite, then each of (a)–(e) is equivalent to

(f) *$S_0(\prec) = \prec$.*

In addition, $(X, \prec)$ is a linearly ordered set if and only if $\sim$ is the identity relation $\{(x, x): x \in X\}$.

Remark 1. We show why (f) can fail to imply weak order when X is not finite. Suppose $(X, \prec)$ consists of the linearly ordered set of integers $(\mathbb{Z}, <)$

plus an additional point ω such that $\omega \sim n$ for all $n \in \mathbb{Z}$. Then $n \prec n+1 \sim \omega \Rightarrow \operatorname{not}(\omega S_0(\prec)n)$, and $\omega \sim (n-1) \prec n \Rightarrow \operatorname{not}(nS_0(\prec)\omega)$. Theorem 1(e) then implies $S_0(\prec) = \prec$, but $\prec$ is not a weak order since $\sim$ is not transitive.

Proof. The proof of the final sentence of Theorem 2 is left to the reader. The proofs of the other implications follow.

(a) $\Rightarrow$ (b). Since negative transitivity says that $x \prec z \Rightarrow (x \prec y$ or $y \prec z)$ and $z \prec x \Rightarrow (z \prec y$ or $y \prec x)$, $x \sim y$ and $y \sim z$ imply $x \sim z$, so $\sim$ is transitive.

(b) $\Rightarrow$ (c). Clearly $x \approx y \Rightarrow x \sim y$, and $x \sim y \Rightarrow (z \sim x \Leftrightarrow z \sim y)$.

(c) $\Rightarrow$ (d). See Theorem 1(b).

(d) $\Rightarrow$ (b). Obvious.

(d) $\Rightarrow$ (e). By Theorem 1(d), $(X/\approx, \prec)$ is a poset. By (d) $\Rightarrow$ (b) $\Rightarrow$ (c), $\approx$ equals $\sim$. To show that $\prec$ on $X/\approx$ is complete, suppose $A, B \in X/\sim$ and $A \neq B$. Then $A \cap B = \varnothing$ since $\sim$ is an equivalence relation by Theorem 1(a). Hence $a \prec b$ or $b \prec a$ for some $(a,b) \in A \times B$. If $a \prec b$ then (d) $\Rightarrow A \prec B$, and if $b \prec a$ then (d) $\Rightarrow B \prec A$.

(e) $\Rightarrow$ (a). Since $(X, \prec)$ is a poset, $\prec$ is asymmetric. To prove that $\prec$ is negatively transitive, let $[x]$ denote the equivalence class in $X/\approx$ that contains x. Suppose $x \prec z$. Then $[x] \cap [z] = \varnothing$. Given any $y \in X$, either $[y] \cap [x] = \varnothing$ or $[y] \cap [z] = \varnothing$. Suppose for definiteness that $[y] \cap [x] = \varnothing$. Then (e) implies either $[x] \prec [y]$ or $[y] \prec [x]$. If $[x] \prec [y]$ then $x \prec y$, and if $[y] \prec [x]$ then $y \prec x$, hence $y \prec z$ by transitivity. Therefore $x \prec z \Rightarrow (x \prec y$ or $y \prec z)$, so $\prec$ is negatively transitive.

(a) $\Rightarrow$ (f). Since (a) $\Rightarrow$ (d), $S_0(\prec) = \prec \cap cd(\prec)$. Since $\prec$ is asymmetric (see preceding paragraph), $x \prec y \Leftrightarrow yd(\prec)x \Rightarrow \operatorname{not}(xd(\prec)y) \Leftrightarrow xcd(\prec)y$, and therefore $\prec \subseteq cd(\prec)$. Hence $S_0(\prec) = \prec$.

[(f) and $|X| < \infty] \Rightarrow$ (a). The conclusion is obvious if $\prec$ is empty. Assume henceforth that $S_0(\prec) = \prec \neq \varnothing$. Let Y be a maximum-cardinality linearly ordered subset of $(X, \prec)$ and let $|Y| = m$. The existence of Y is guaranteed by the finiteness of X. Let

$$A_k = \{x\colon \text{the maximum-cardinality chain in which } x \text{ is the final element has } k \text{ elements}\}.$$

Clearly, $x \sim y$ when $x, y \in A_k$. This and the transitivity of $\prec$ show that $(x \in A_j, y \in A_k, j < k) \Rightarrow \operatorname{not}(y \prec x)$. Hence to prove that $\prec$ is a weak order, we need only show that $A_1 \prec A_2 \prec \cdots \prec A_m$. Suppose this is false. Let k be the smallest integer for which $\operatorname{not}(A_{k-1} \prec A_k)$. Then $x_{k-1} \sim x_j$ for some $x_{k-1} \in A_{k-1}$ and $x_j \in A_j$ with $j \geq k$. Choose x_j for this so that j is as large as possible. Then $x_{k-1}(\sim)(\prec)x_j$ since, by the definition of A_j, $y_{k-1} \prec x_j$ for some $y_{k-1} \in A_{k-1}$. If $z \prec x_{k-1}$ then, by the choice of k, $z \prec y_{k-1}$, hence $z \prec x_j$, hence $\operatorname{not}(x_j \sim z \prec x_{k-1})$, so that $(x_{k-1}, x_j) \in cd((\sim)(\prec))$. Moreover, $(x_{k-1}, x_j) \in cd((\prec)(\sim))$ since if $x_j(\prec)(\sim)x_{k-1}$ then $x_j \prec v \sim x_{k-1}$ for some v, which contradicts the choice of j. Since $cd((\sim)(\prec)) \cap cd((\prec)(\sim)) =$

$cd((\sim)(\prec) \cup (\prec)(\sim))$, $x_{k-1}S_0(\prec)x_j$, and therefore $x_{k-1} \prec x_j$ by (f). But this contradicts our hypothesis that $x_{k-1} \sim x_j$, and we conclude that $A_1 \prec A_2 \prec \cdots \prec A_m$. □

1.4 LINEAR EXTENSIONS

A *linear extension* of a poset $(X, \prec)$ is a linearly ordered set $(X, \prec^*)$ for which $\prec \subseteq \prec^*$. This section first proves the extension theorem of Szpilrajn (1930), then uses his theorem in the proof of a numerical representation theorem for posets that have countably many $\approx$ classes.

The proof of Szpilrajn's theorem uses

Kuratowski's Lemma. *Each chain in a poset is included in a maximal chain of the poset.*

Kelley (1955, p. 33) notes that this is equivalent to several other set-theoretic axioms, including the Axiom of Choice—if $\mathscr{F}$ is a set of nonempty sets, then there is a function g on $\mathscr{F}$ such that $g(A) \in A$ for every $A \in \mathscr{F}$.

Theorem 3 (Szpilrajn). *Every poset has a linear extension.*

Proof. Let $(X, \prec)$ be a poset. Assume that $\prec$ is not complete since otherwise $(X, \prec)$ is a chain. Let $\sim = sc(\prec)$, take $x \sim y$ for $x \neq y$, and define $\prec'$ on X by

$$\prec' = \prec \cup [\{a: a \prec x \text{ or } a = x\} \times \{b: y \prec b \text{ or } y = b\}].$$

It is routine to verify that $(X, \prec')$ is a poset. Moreover, $\prec \subset \prec'$ since $x \prec' y$.

Let $\mathscr{P}$ be the set of all partial orders P on X such that $\prec \subseteq P$, and order $\mathscr{P}$ by inclusion so that $(\mathscr{P}, \subset)$ is a poset. Let $(\mathscr{P}', \subset)$ be any nonempty chain in $(\mathscr{P}, \subset)$. Then, by Kuratowski's lemma, there is a maximal chain $(\mathscr{P}^*, \subset)$ in $(\mathscr{P}, \subset)$ that includes $(\mathscr{P}', \subset)$. Given such a $(\mathscr{P}^*, \subset)$, let

$$\prec^* = \bigcup \{P: P \in \mathscr{P}^*\}.$$

Since $\mathscr{P}^*$ is linearly ordered by $\subset$, it is easily seen that $\prec^*$ is a partial order. Suppose $\prec^*$ is not complete. Then, by the preceding paragraph, there is a partial order $\prec^{*\prime}$ on X that properly includes $\prec^*$. But then $(\mathscr{P}^*, \subset)$ is not maximal since $P \subset \prec^{*\prime}$ for every $P \in \mathscr{P}^*$, and we obtain a contradiction. Therefore $\prec^*$ is complete, so it gives a linear extension of $\prec$. □

Since x and y can be interchanged in the first paragraph of the preceding proof, it follows that for every $(x, y) \in sc(\prec)$ for which $x \neq y$, there are linear extensions of $(X, \prec)$ in which x precedes y, and others in which y precedes x. Consequently, the intersection of all linear extensions of a poset equals the poset. The minimal cardinality over the sets of linear extensions whose intersections equal the poset is called the *dimension* of the poset. We shall return to this in Chapter 5.

We now use Szpilrajn's theorem to prove

Theorem 4. *Suppose $(X, \prec)$ is a poset for which $X/\approx$ is countable when $\approx$ is defined as in the preceding section. Then there exists $f: X \to \mathbb{R}$ such that, for all $x, y \in X$,*

$$x \approx y \Leftrightarrow f(x) = f(y),$$

$$x \prec y \Rightarrow f(x) < f(y).$$

Remark 2. Theorems 2 and 4 imply that if $(X, \prec)$ is a weakly ordered set such that $X/sc(\prec)$ is countable, then there exists $f: X \to \mathbb{R}$ such that, for all $x, y \in X$,

$$x \prec y \Leftrightarrow f(x) < f(y).$$

This is true also if $(X, \prec)$ is a countable chain, but may be false when X is uncountable. The latter case is discussed further in Chapter 7.

Proof. Given the hypotheses of Theorem 4, Theorems 1(d) and 3 imply that there is a linear order $\prec^*$ on $X/\approx$ that includes $\prec$ on $X/\approx$. Since $X/\approx$ is countable, it can be enumerated as $A_1, A_2, \ldots$. Define $F: X/\approx \to \mathbb{R}$ by

$$F(A_k) = \Sigma\{2^{-j}: A_j \prec^* A_k\},$$

where the summation is over those j for which $A_j \prec^* A_k$. If $A_i \prec^* A_k$ then $\{j: A_j \prec^* A_i\} \subset \{j: A_j \prec^* A_k\}$ so that $F(A_i) < F(A_k)$. Since $\prec^*$ is a chain, all F values are distinct. Moreover, if $A, B \in X/\approx$ and $A \prec B$, then $A \prec^* B$, so $F(A) < F(B)$. The conclusion of the theorem follows on defining $f(x)$ as $F(A)$ for all $x \in A$ and all $A \in X/\approx$. □

1.5 TRANSITIVE REORIENTATIONS

A *reorientation* of an asymmetric related set (X, R) is an asymmetric related set (X, S) for which

$$S \cup d(S) = R \cup d(R).$$

Thus $xRy \Rightarrow (xSy \text{ or } ySx)$, and $xSy \Rightarrow (xRy \text{ or } yRx)$. A reorientation (X, S)

of an asymmetric related set (X, R) is *transitive* if S is transitive, that is, if (X, S) is a poset.

In this section we shall prove an intriguing result of Ghouilà-Houri (1962) which says that an asymmetric related set (X, R) has a transitive reorientation if it satisfies the pseudo-transitivity property $R^2 \subseteq R \cup d(R)$, that is, if for all x, y, and z in X,

$$(xRy, yRz) \Rightarrow (xRz \text{ or } zRx).$$

The proof for finite X follows Ghouilà-Houri (1962) and Berge (1973, pp. 365–366). The proof for infinite X is suggested by Wolk (1965) on the basis of Rado's theorem (Rado, 1949) which, as noted by Mirski and Perfect (1966), has relevance for diverse problems.

I state Rado's theorem without proof: see Mirski and Perfect (1966, p. 540) for references to several proofs. A *choice function* g on a set $\mathcal{F}$ of nonempty sets is a function g on $\mathcal{F}$ such that $g(A) \in A$ for every $A \in \mathcal{F}$. (Cf. Axiom of Choice in the preceding section.)

Rado's Theorem. *Suppose J is a nonempty set, $\mathcal{F} = \{X_j : j \in J\}$ is a family of nonempty finite sets (one for each j, with or without duplications) and, for every finite $A \subseteq J$, g_A is a choice function on $\{X_j : j \in A\}$. Then there is a choice function g on $\mathcal{F}$ such that, for every finite $A \subseteq J$ there is a finite $B \subseteq J$ for which $A \subseteq B$ and $g(X_i) = g_B(X_i)$ for all $i \in A$.*

If J is finite then the theorem is trivial: just take $g = g_J$. Its general form presumes the Axiom of Choice since it presumes the existence of a choice function g on $\mathcal{F}$ regardless of the nature of J.

Theorem 5 (Ghouilà-Houri). *An asymmetric related set (X, R) has a transitive reorientation if $R^2 \subseteq R \cup d(R)$.*

Proof. Assume throughout that (X, R) is asymmetric with $R^2 \subseteq R \cup d(R)$. Call x and y *adjacent* if xRy or yRx, and observe that if x, y, and z form an R-cycle, say $(x, y), (y, z), (z, x) \in R$, then every other point in X is adjacent to 0, 2, or 3 of x, y, and z.

We consider finite X first and proceed by induction on $|X|$, noting that each nonempty restriction of (X, R) inherits the properties assumed for (X, R). The theorem is clearly true when $|X| = 1$. Suppose $|X| = n > 1$ and the theorem holds for sets with fewer than n points. Suppose in addition that R is not transitive, since otherwise there is nothing to prove. Then there exist x_1, x_2, and x_3 in X that cycle in R:

$$x_1 R x_2, x_2 R x_3, x_3 R x_1.$$

Two cases require analysis.

CASE 1. *Each $x \in X \setminus \{x_1, x_2, x_3\}$ that is adjacent to at least one of x_1, x_2, and x_3 is adjacent to all three.* Let $(X \setminus \{x_2, x_3\}, \prec)$ be a transitive reorientation of $(X \setminus \{x_2, x_3\}, R)$, as assured by the induction hypothesis. For each x not in $\{x_1, x_2, x_3\}$ that is adjacent to these three, augment $\prec$ by taking

$$x \prec x_2 \quad \text{and} \quad x \prec x_3 \quad \text{if } x \prec x_1;$$

$$x_2 \prec x \quad \text{and} \quad x_3 \prec x \quad \text{if } x_1 \prec x.$$

Also take $\{x_1 \prec x_2, x_2 \prec x_3, x_1 \prec x_3\}$. Then $(X, \prec)$ is transitive: see the left part of Fig. 1.2.

CASE 2. *There is a $y \in X \setminus \{x_1, x_2, x_3\}$ that is adjacent to exactly two x_i, say x_2 and x_3.* Let

$$A = \{z \in X : zRx_2 \text{ and } x_3 Rz\},$$

so that each point z in A is in a cycle $zRx_2 Rx_3 Rz$. Clearly $x_1 \in A$, and $y \in A$ also since y is not adjacent to x_1.

We claim that *if $x \notin A$, then either x is adjacent to everything in A or to nothing in A.* To the contrary, suppose x is not in A, a and b are in A, x is adjacent to a, and x is not adjacent to b. Then x is adjacent to at least two points in $\{a, x_2, x_3\}$ and is therefore adjacent to x_2 or x_3. We see from $\{b, x_2, x_3\}$ that x is adjacent to both x_2 and x_3 since it is not adjacent to b. Moreover, xRx_2 and $x_3 Rx$. But this contradicts $x \notin A$.

Let $\prec_1$ and $\prec_2$ respectively be transitive reorientations of the restrictions of R to A and to $(X \setminus A) \cup \{x_1\}$, as assured by the induction hypothesis. In addition, for each $x \notin A$ that is adjacent to some (hence to all) $a \in A$, define $\prec_3$ on pairs $\{x, a\}$ by

$$x \prec_3 a \quad \text{if } x \prec_2 x_1,$$

$$a \prec_3 x \quad \text{if } x_1 \prec_2 x.$$

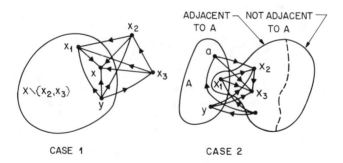

CASE 1   CASE 2

Figure 1.2

Then $\prec_1 \cup \prec_2 \cup \prec_3$ is a transitive reorientation of R on X: see the right part of Fig. 1.2.

This completes the proof of Theorem 5 for finite X. Suppose now that X is infinite, and that R is nonempty. Let

$$J = \{\{x, y\}: xRy \text{ or } yRx\},$$

$$X_{\{x,y\}} = \{(x, y), (y, x)\}.$$

Given nonempty finite $A \subseteq J$, let X_A be the union of the $\{x, y\}$ in A. By the preceding proof, there is a poset $(X_A, \prec_A)$ with $\prec_A \cup d(\prec_A)$ equal to the restriction of $R \cup d(R)$ on A. For each such A define g_A on $\{X_{\{x,y\}}: \{x, y\} \in A\}$ by

$$g_A(\{x, y\}) = (x, y) \quad \text{if } x \prec_A y.$$

Note that if $\{x, y\}, \{y, z\} \in A$ with $g(\{x, y\}) = (x, y)$ and $g_A(\{y, z\}) = (y, z)$, then the construction implies $x \prec_A z$, hence that $\{x, z\}$ is in $R \cup d(R)$ regardless of whether it is in A.

Let g be a choice function on $\{X_{\{x,y\}}: \{x, y\} \in J\}$ of the type guaranteed by Rado's theorem, and define $\prec$ on X by

$$x \prec y \text{ if } \{x, y\} \in J \quad \text{and} \quad g(\{x, y\}) = (x, y).$$

Then $\prec$ is obviously asymmetric. To verify transitivity, suppose $x \prec y$ and $y \prec z$. Let $A = \{\{x, y\}, \{y, z\}\}$. Then there is a finite $B \subseteq J$ that includes A with $g_B = g$ on A. In particular, $g_B(\{x, y\}) = g(\{x, y\}) = (x, y)$ and $g_B(\{y, z\}) = g(\{y, z\}) = (y, z)$. By the final sentence of the preceding paragraph, $\{x, z\} \in J$, so $\{\{x, y\}, \{y, z\}, \{x, z\}\}$ is a subset of J and it has a finite superset $C \subseteq J$ with $g_C = g$ on $\{\{x, y\}, \{y, z\}, \{x, z\}\}$. It follows that $g(\{x, z\}) = g_C(\{x, z\}) = (x, z)$, hence that $x \prec z$. □

We shall use Theorem 5 in Section 1.7 to complete the proof of a characterization of comparability graphs.

1.6 INTERVALS AND ANTICHAINS

Throughout this section, $\sim\, = sc(\prec)$ and $\approx$ is the poset equivalence relation defined in Section 1.3. We shall also define $\precsim$ and $\succsim$ by

$$x \precsim y \quad \text{if } x \prec y \quad \text{or} \quad x \sim y,$$

$$x \succsim y \quad \text{if } x \prec y \quad \text{or} \quad x \approx y.$$

An *interval* in a poset $(X, \prec)$ is a nonempty subset Y of X such that, for all $x, y, z \in X$,

$$(x, y \in Y, x \prec z \prec y) \Rightarrow z \in Y.$$

An interval is *inclusive* if it contains every x that is equivalent to a point in the interval, that is, if

$$(x, y \in Y, x \precsim z \precsim y) \Rightarrow z \in Y.$$

Thus, an inclusive interval is the union of equivalence classes. The latter category is more restricted than the first but is still very broad since, for example, every poset is itself an inclusive interval.

More familiar notions of intervals obtain for weakly ordered sets since then $X/\sim$ is linearly ordered by $\prec$ (Theorem 2). An interval Y in a weakly ordered set is *closed* if there exist $x, y \in Y$ such that $Y \subseteq \{z: x \precsim z \precsim y\}$ and is *open* if for every $z \in Y$ there are $x, y \in Y$ such that $x \prec z \prec y$.

Because $\approx$ is the identity relation when $(X, \prec)$ is a linearly ordered set, all intervals are inclusive intervals in this case.

An *antichain* in a poset $(X, \prec)$ is an interval A for which $(A \times A) \cap \prec = \emptyset$. Equivalently, A is an antichain if $A \times A \subseteq sc(\prec)$, that is, $x \sim y$ for all $x, y \in A$. Every singleton subset of a poset is both a chain and an antichain, but no subset with two or more elements can be both. With respect to a given poset, a *maximal* antichain is an antichain not properly included in another antichain, and a *maximum* antichain (if it exists) is a maximum-cardinality antichain.

There are interesting relationships between chains and antichains in posets observed by Dilworth (1950) and others. We mention two of these here.

Theorem 6 (Dilworth). *Suppose the maximum antichain in a poset $(X, \prec)$ has $m \in \{1, 2, \ldots\}$ elements. Then there are m chains in $(X, \prec)$, say $(X_1, \prec_1), \ldots, (X_m, \prec_m)$, such that $X = \bigcup_{i=1}^{m} X_i$.*

Proof. We give a proof only for finite X using induction on $|X|$ in the manner of Perles (1963a) and Tverberg (1967). Rado's theorem can be used for the infinite case.

The theorem is clearly true when $|X| = 1$. Assume henceforth that the theorem is true for posets smaller than $|X| = n \geq 2$, let $(Y, \prec)$ be a maximum chain in the poset $(X, \prec)$, and let y^* be the maximal element in Y. If no antichain in $(X \setminus Y, \prec)$ has m elements, then the induction hypothesis implies that $X \setminus Y$ is the union (of the X_i) of $m - 1$ chains, so X is the union of m chains.

Assume henceforth that $(X \setminus Y, \prec)$ includes maximum antichain $A = \{a_1, \ldots, a_m\}$. Let

$$X^- = \{x \in X: x \precsim a_i \text{ for some } i\},$$

$$X^+ = \{x \in X: a_i \precsim x \text{ for some } i\}.$$

If $y^* \in X^-$, then $y^* \prec a_i$ for some i, contrary to the maximality of $(Y, \prec)$. Hence $y^* \notin X^-$, and therefore, by the induction hypothesis, X^- is the union (of the X_i^-) of m chains $(X_1^-, \prec_1^-), \ldots, (X_m^-, \prec_m^-)$, where $a_i \in X_i^-$ and $\prec_i^- = \prec \cap (X_i^- \times X_i^-)$ for each i. If $x \in X_i^-$, then $x \leq a_j$ for some j, and if $a_i \prec x$ then $a_i \prec a_j$, a contradiction. Therefore a_i is the maximal element in X_i^-.

By a symmetric argument, X^+ is the union of m chains $(X_1^+, \prec_1^+), \ldots, (X_m^+, \prec_m^+)$ with $a_i \in X_i^+$, $\prec_i^+ = \prec \cap (X_i^+ \times X_i^+)$ and a_i minimal in X_i^+ for each i. It follows that $(X_i^- \cup X_i^+, t(\prec_i^- \cup \prec_i^+))$ is a chain in $(X, \prec)$ for each i with $X = \cup_i (X_i^- \cup X_i^+)$. □

Figure 1.3 illustrates Theorem 6 with the Hasse diagram of a poset on 13 points. The maximum antichains have three points, and the 13 points are covered by three chains, labeled 1, 2, and 3. These chains do not involve edges e and e', and it is easily seen that five chains are needed to cover all the edges.

The question of how many chains are needed to cover all the edges in a diagram of a finite poset is answered by Bogart (1970). His theorem says that if there are k edges that are mutually incomparable in the sense that one of the two points in one edge bears $\sim$ to a distinct point in the other edge, but this is not true for any $k + 1$ edges, then k chains (and no fewer) are needed to cover all the edges. One such $k = 5$ set is identified by the dashed line in Fig. 1.3. Bogart's proof uses Dilworth's theorem.

1.7 GRAPHS

A *graph* is a symmetric related set $(X, \sim)$. In the present section, we do not assume any relationship between $\sim$ and $sc(\prec)$ unless noted otherwise, but here and elsewhere $\sim$ is always presumed to be symmetric.

A graph $(X, \sim)$ is *reflexive* if $\sim$ is reflexive (in which case it has a *loop* (x, x) at each point), *irreflexive* if $\sim$ is irreflexive (loopless, as often assumed in works on graph theory), and *complete* if $\sim$ is complete. A complete irreflexive graph on n points is denoted by K_n. The same notation is sometimes used for a complete reflexive graph on n points, and I shall do so later.

Because $\sim$ is symmetric, $a(X, \sim) = (X, a(\sim)) = (X, \emptyset)$, $s(X, \sim) = (X, \sim)$, and $d(X, \sim) = (X, \sim)$. The complement $c(X, \sim) = (X, c(\sim))$ of a

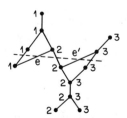

Figure 1.3

graph never equals the graph, $c(c(X, \sim)) = (X, \sim)$, and a graph is reflexive if and only if its complement is irreflexive. As defined below, the components of a graph are the maximal complete subgraphs of its transitive closure $t(X, \sim)$.

The restriction $(Y, \sim \cap (Y \times Y))$ of a graph $(X, \sim)$ on a nonempty subset Y of X is the *subgraph* (induced) on Y. A complete subgraph is also called a *clique*, and a subgraph $(Y, \sim)$ for which $x \not\sim y$ for all distinct x and y in Y is an *independent set*. A graph $(X, \sim)$ is *n-partite* if X can be partitioned into n independent sets. A 2-partite graph is usually said to be *bipartite*. The standard usages of maximal and maximum apply to cliques and independent sets.

A *path* in $(X, \sim)$ is a sequence $(x_0, x_1, \ldots, x_m)$ that has $x_i \sim x_{i+1}$ for all $0 \leq i < m$. It has *length m* (one less than the number of terms), and is *simple* if all x_i are distinct. A *cycle* is a path $(x_0, x_1, \ldots, x_{m-1}, x_0)$, so that $x_m = x_0$. It has length m, and is *simple* if x_0 through x_{m-1} are distinct. A standard result in graph theory says that an irreflexive graph is bipartite if and only if it has no cycle of odd length. A *chord* of cycle $(x_0, x_1, \ldots, x_{m-1}, x_0)$ is an (x_i, x_j) in $\sim$ for which the absolute difference between i and j is between 2 and $m - 2$ inclusive. A chord of the form (x_i, x_{i+2}), (x_{m-2}, x_0), or (x_{m-1}, x_1) is a *triangular chord*.

Distinct points x and y in $(X, \sim)$ are *connected* if $xt(\sim)y$. This holds if and only if there is a simple path $(x, \ldots, y)$ in the graph. A graph is *connected* if every two distinct points are connected. A *component* of $(X, \sim)$ is a maximal connected subgraph of $(X, \sim)$, that is, a maximal clique of $(X, t(\sim))$.

A point x in $(X, \sim)$ is *isolated* if $y \not\sim x$ for every $y \neq x$ in X, and in this case $\{x\}$ is a component of the graph. A point u is *universal* if $u \sim X$. If $(X, \sim)$ has a universal point, then it is obviously connected. If $(X, \sim)$ is not complete and has the set U of universal points, then it is easily seen that the subgraph on $X \setminus U$ has no universal point.

Several specialized types of graphs are important in the study of ordered sets. Paramount among these are irreflexive graphs whose "edges" $\{(x, y), (y, x)\}$ can be "directed" by the choice of one of (x, y) and (y, x) for each $x \sim y$ so that the set of choices is a specified type of order relation. Thus, the

QUESTION: What conditions on an irreflexive graph $(X, \sim)$ are necessary and sufficient for the existence of an irreflexive order (of a specified type) R on X for which $\sim$ equals $R \cup d(R)$?

The answer is obvious for a linear order R: $(X, \sim)$ must be a complete graph. When R is a weak order, it is necessary and sufficient for $sc(\sim)$ to be an equivalence relation. The classes in $X/sc(\sim)$ then constitute the classes in $X/sc(R)$.

The answer to the Question when R is an interval order is given in Chapter 3, but a warning is needed at this point. What we shall refer to as an interval graph $(X, \sim)$ in Chapter 3 and thereafter is *not* the $(X, \sim)$ of the Question, but is rather the *complement* of the $(X, \sim)$ of the Question. In other words, an irreflexive graph of the Question can have its edges directed to yield an interval order if, and only if, its complement is an interval graph.

A graph $(X, \sim)$ for which there exists a *partial order* $\prec$ on X such that $\sim \; = \; \prec \cup d(\prec)$ is called a *comparability graph*. The following important theorem was initially proved by Gilmore and Hoffman (1964) and Ghouilà-Houri (1962). The proof given below is patterned after Ghouilà-Houri (1962) and Berge (1973).

Theorem 7 (Gilmore and Hoffman; Ghouilà-Houri). *A graph $(X, \sim)$ is a comparability graph if and only if it is irreflexive and every cycle of odd length has a triangular chord.*

Proof. Necessity. The necessity of irreflexivity is obvious. Given irreflexivity, if m is odd and cycle $(x_0, x_1, \ldots, x_{m-1}, x_0)$ has no triangular chord, then the desired transitivity of the agreeing partial order forces $x_0 \prec x_1, x_2 \prec x_1$, $x_3 \prec x_2, \ldots, x_{m-2} \prec x_{m-1}, x_{m-1} \prec x_0$, and then $x_1 \prec x_0$ (or the duals of these), which violates asymmetry.

Sufficiency. Assume that $(X, \sim)$ is irreflexive and every odd cycle has a triangular chord. Let $(\sim, I)$ be the graph whose points are the ordered pairs in $\sim$ (assume nonempty) such that, for all $\alpha, \beta \in \sim$,

$$\alpha I \beta \quad \text{if} \quad \{\alpha, \beta\} = \{(x, y), (y, z)\} \text{ for some } x, y, \text{ and } z \text{ in } X, \text{ and } x \not\sim z.$$

If $x \sim y$, then $(x, y) I(y, x)$ since $\sim$ is irreflexive.

We claim that $(\sim, I)$ has no cycle of odd length. To the contrary, suppose $\alpha_0 I \alpha_1 I \cdots I \alpha_{m-1} I \alpha_0$ with m odd. Increase this cycle by additions as follows for $i = 0, \ldots, m - 1$: if $\alpha_i = (x, y)$ and $\alpha_{i+1} = (z, x)$ with $z \neq y$, replace $\alpha_i I \alpha_{i+1}$ by $\alpha_i I(y, x) I(x, z) I \alpha_{i+1}$. The augmented I cycle has odd length since each addition adds two terms, and it has the form $(x_0, x_1) I(x_1, x_2) I(x_2, x_3) I \cdots I(x_{n-1}, x_0) I(x_0, x_1)$ with n odd. The corresponding $\sim$ cycle, namely $x_0 \sim x_1 \sim x_2 \sim \cdots \sim x_{n-1} \sim x_0$, has length n and, by the definition of I, it has no triangular chord. But this contradicts our assumptions.

Since $(\sim, I)$ has no odd cycle, it is a bipartite graph. Therefore $\sim$ can be partitioned into $\sim_1$ and $\sim_2$ such that, whenever α and β are in the same $\sim_i$, $(\alpha, \beta) \notin I$. Given such a partition, define $\prec_1$ on X by

$$x \prec_1 y \quad \text{if} \quad x \sim_1 y,$$

and note that $\sim_2 \; = \; d(\sim_1)$, that is, $x \sim_2 y \Leftrightarrow y \sim_1 x$. In particular, $\prec_1$ is asymmetric and $\sim \; = \; \prec_1 \cup d(\prec_1)$. Furthermore,

$$(x \prec_1 y, y \prec_1 z) \Rightarrow [(x, y), (y, z) \in \sim_1] \Rightarrow ((x, y), (y, z)) \notin I$$
$$\Rightarrow x \sim z \Rightarrow (x \prec_1 z \text{ or } z \prec_1 x),$$

so that $(\prec_1)^2 \subseteq \prec_1 \cup d(\prec_1)$.

It follows from Theorem 5 that $(X, \prec_1)$ has a transitive reorientation $(X, \prec)$. Since $\sim \, = \, \prec \cup \, d(\prec)$, we have shown that $(X, \sim)$ is a comparability graph. □

1.8 PREVIEW

Our study of interval orders begins in the next chapter with simple axiomatic definitions of interval orders and semiorders, which are special types of interval orders. The basic structures of these ordering relations are developed, and representation theorems for mappings into intervals of linearly ordered sets are established. Real interval representations for finite and countable cases are also noted. The chapter concludes with comments on extending posets and interval orders to weakly ordered sets.

Interval graphs and indifference graphs, which correspond respectively to interval orders and semiorders, are defined in Chapter 3 through intervals of linearly ordered sets. Traditional characterizations of these graphs in terms of comparability graphs and related concepts are then presented. Further relationships between interval orders and interval graphs are noted along with two characterizations of interval graphs that have essentially unique agreeing interval orders. One of these characterizations is based on Hanlon's (1982) concept of buried subgraph, which will play a role in later chapters.

Chapter 4 departs from the main flow of the book to identify conditions on a ternary betweenness relation T on X that are necessary and sufficient for the existence of an ordering relation $\prec$ on X that agrees with T in the sense that

$$(x, y, z) \in T \Leftrightarrow (x \prec y \prec z \text{ or } z \prec y \prec x).$$

We shall do this for interval orders, semiorders, weak orders, and linear orders, and will note several related results. The background of this work stems from investigations of E. V. Huntington and others on betweenness characterizations of linearly ordered sets earlier in this century.

Chapter 5 returns to our main concern with interval orders and interval graphs in the binary mode. It describes various features or parameters of posets, including dimensionality, height, width, breadth, and so on. We then concentrate on dimensionality and establish several results for posets, semiorders, and interval orders.

The focus of Chapter 6 is the maximum semiorder included in an interval order, or the maximum indifference graph included in an interval graph. Although a number of results are proved about the function $\tau(n)$, defined as the largest k such that every interval order [interval graph] on n points includes a k-point semiorder [indifference graph], important questions about τ remain open.

Chapter 7 returns to the theme of representations by real intervals begun in Chapter 2. Its initial aim is to prove the Scott-Suppes (1958) single-length theorem for finite semiorders by two methods. The first is a constructive counting procedure; the second, which is used for more complex theorems in the next chapter, is a nonconstructive linear separation approach. Chapter 7 then develops real interval representations for sets of arbitrary cardinality on the basis of Cantor's representation theorem for linearly ordered sets.

Chapter 8 expands the theme of single-length representations for finite semiorders and indifference graphs by introducing classes of finite interval orders and interval graphs that have representations whose interval lengths are bounded between two positive numbers. Classes with rational bounds are characterized by axioms for compositions of $\prec$ and its symmetric complement $\sim$. Example: A finite interval order has a representation whose interval lengths are all in $[1, 2]$ if and only if $(\prec)^3(\sim) \subseteq \prec$.

In contrast to bounds on interval lengths, Chapter 9 considers classes of finite interval orders and interval graphs that have representations which use no more than n distinct interval lengths. It is shown that the $n = 2$ case cannot be characterized by a finite set of forbidden interval orders or subgraphs. The same thing is true for larger n. Examples illustrate several paradoxical possibilities for $n = 2$ and other cases.

The final chapter, like Chapter 6, focuses on an extremization problem. In Chapter 10 we investigate the problem of determining, for each $k \geq 2$, the smallest integer n such that there is an interval order on n points whose representation requires at least k lengths. Again, a number of results are proved, but the general problem and several important auxiliary problems remain open.

2
Interval Orders

This chapter defines interval orders, partial semiorders, and semiorders by axioms for $\prec$ on X and develops their basic structures. The underlying structures of interval orders and semiorders are then used to prove representation theorems for mappings into intervals of linearly ordered sets. The final section discusses extensions to weakly ordered sets.

2.1 DEFINITIONS AND EXAMPLES

An *interval order* is an irreflexive related set $(X, \prec)$ that satisfies

$$(a \prec x, b \prec y) \Rightarrow (a \prec y \text{ or } b \prec x), \quad \text{for all } a, b, x, y \in X. \qquad (1)$$

A *partial semiorder* is an irreflexive related set $(X, \prec)$ that satisfies

$$(a \prec b, b \prec c) \Rightarrow (a \prec x \text{ or } x \prec c), \quad \text{for all } a, b, c, x \in X. \qquad (2)$$

A *semiorder* is an irreflexive related set that satisfies both (1) and (2). When $(X, \prec)$ is an interval order, partial semiorder, or semiorder, $\prec$ itself will sometimes be referred to by the same designation.

Property (2), which is tantamount to $(\prec)^2 \subseteq c(c(\prec)c(\prec))$, is sometimes referred to as semitransitivity (Chipman, 1971), and a partial semiorder is sometimes called a semitransitive order.

The use of "order" in each of the foregoing definitions refers to the fact that *every interval order and every partial semiorder is transitive* and is therefore a poset. For, if $x \prec y$ and $y \prec z$, then (1) yields $x \prec z$ or $y \prec y$, and (2) yields $x \prec z$ or $z \prec z$, so irreflexivity forces $x \prec z$ in both cases.

The smallest poset that is not also an interval order has four points. The canonical case is shown by the Hasse diagram I in Fig. 2.1. According to (1), a poset is an interval order if and only if it has no restriction that is isomorphic to the poset of diagram I.

The smallest poset that is not also a partial semiorder has four points. The canonical case is shown by diagram II in Fig. 2.1. By (2), a poset is a partial semiorder if and only if it has no restriction that is isomorphic to the poset of diagram II.

Definitions and Examples

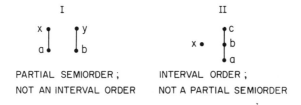

PARTIAL SEMIORDER ; INTERVAL ORDER ;
NOT AN INTERVAL ORDER   NOT A PARTIAL SEMIORDER

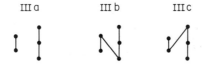

POSETS THAT ARE NEITHER INTERVAL
ORDERS NOR PARTIAL SEMIORDERS

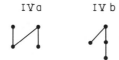

SEMIORDERS THAT ARE NOT WEAK ORDERS

Figure 2.1 Hasse diagrams of posets.

The smallest X that admits a partial order that is neither an interval order nor a partial semiorder has five points: see Fig. 2.1 (IIIa–IIIc). Diagrams IVa and IVb illustrate semiorders that are not weak orders.

Throughout this chapter we shall let $\sim$ denote the symmetric complement of $\prec$ and let $\approx$ be the equivalence relation of Section 1.3:

$$x \sim y \text{ if not}(x \prec y) \text{ and not}(y \prec x);$$
$$x \approx y \text{ if } \{z: z \sim x\} = \{z: z \sim y\}.$$

The relation $\sim$ is nontransitive for every poset in Fig. 2.1. Thus, by Theorem 1.2(a, b), none of the posets in the figure is a weakly ordered set.

In his 1914 paper, Wiener recognized interval orders as an important concept for the analysis of temporal events, each of which occurs over some time span. If we say that event x precedes event y when x ends before y begins, then precedence is an interval order on a set of events. Specific examples arise in chronological dating in archaeology, spans for the existence of species in paleontology, production schedules, and orderings of atomic events.

More generally, Wiener noted that the concepts of precedence and interval orders apply to comparisons of intervals in any linearly ordered set. As an example, consider ordinary measurement of length or mass, or some other physical property, that is subject to error. Then each measurement is given as an interval rather than a single point, and, within the accuracy of the measuring device, notions of "longer than," "heavier than," and so forth, are interval orders rather than weak orders. A similar example arises from rounding error in digital computation.

Interest in interval orders and semiorders in the behavioral sciences arose from several sources. One is Keynes's treatise on probability (1921). From the present perspective, it may be more reasonable to represent individuals' subjective probabilities for uncertain events by intervals in [0, 1] rather than by single numbers to allow for uncertainties in judgment. For example, if A = "rain will fall on New York City next June 15," B = "snow will fall on Chicago next March 1," C = "snow will fall on Chicago next March 2," and if $\prec$ means "is less probable than," then $\{A \sim B, A \sim C, C \prec B\}$ is a reasonable set of judgments. In this simple case, $(\{A, B, C\}, \prec)$ is a semiorder but not a weak order since $\sim$ is not transitive.

Armstrong's work (1939, 1948, 1950) on preference comparisons in consumer economics introduced the notion of semiorders into economic theory although the name "semiorder" was first used and precisely axiomatized by Luce (1956). When $\prec$ is interpreted as "less preferred than," or its dual $\succ$ as "preferred to," $\sim$ is called an *indifference* relation. Armstrong (1950, p. 122) speaks of nontransitive indifference as arising from "the imperfect powers of discrimination of the human mind whereby inequalities become recognizable only when of sufficient magnitude." To use an example of Luce's, it is reasonable to suppose that a person will be indifferent between x and $x + 1$ grains of sugar in his coffee for $x = 0, 1, \ldots$, yet have a definite preference between 0 grains and 5000 grains. An example of the depth to which intransitive indifference has been probed in economic theory is Chipman's (1971) analysis of partial semiorders as preference relations on spaces of commodity bundles.

Psychophysical measurement of perceptions of length, pitch, loudness, and so forth, provides other examples of qualitative comparisons that might be analyzed from the perspective of interval orders and semiorders rather than the more precise but less realistic weak orders and linear orders. Interest in (2) as well as (1) arose in part from the notion of thresholds of discrimination or just-noticeable-differences (jnd's) in psychophysical judgment (Luce and Suppes, 1965). In psychophysical measurement, $x \prec y$ indicates that there is a noticeable difference in a particular direction between x and y, which are therefore separated by at least one jnd. If $a \prec b$ and $b \prec c$ then a and c are separated by at least two jnd's. Hence, if $a \prec b \prec c$ and if x is any other stimulus, then either there is a jnd between a and x or a jnd between x and c, and it is natural to propose the axiom

$$(a \prec b, b \prec c) \Rightarrow (\text{not}(a \sim x) \text{ or } \text{not}(x \sim c)).$$

Basic Structures 21

This is easily seen to be equivalent to (2) when irreflexivity and (1) are assumed.

The failure of (2) in the psychophysical setting could indicate that discriminatory thresholds vary depending on the stimuli being compared. For example, if b is a relatively pure tone but x is clouded by disturbances such as oscillating volume, and if $\prec$ refers to judgments of pitch, then $a \prec b \prec c$ along with $a \sim x \sim c$ might obtain.

2.2 BASIC STRUCTURES

We first note a few facts about posets and partial semiorders before taking a closer look at interval orders. In addition to the things proved in Chapter 1 for posets, we mention incidentally that a partial order $\prec$ on X is the disjoint union of its *diagram* $\prec \setminus \prec^2$ (as in Hasse diagram) and $\prec^2$, which is also a partial order. A similar result applies to partial semiorders, with a slight twist on $\prec^2$. As a definition, we say that $(X, \prec)$ is *dense* if its diagram is empty, that is, $\prec \subseteq \prec^2$ or, whenever $x \prec y$, then $x \prec z \prec y$ for some z in X.

Theorem 1. *If $(X, \prec)$ is a partial semiorder, then $\prec^2$ is a semiorder. Moreover, a dense partial semiorder is a semiorder.*

Proof. Let $(X, \prec)$ be a partial semiorder. Then $\prec^2$ is clearly irreflexive. To prove that (2) holds for $\prec^2$, suppose that $a \prec^2 b$ and $b \prec^2 c$, and let x be any point in X. For definiteness take $a \prec y \prec b$ and $b \prec z \prec c$. Then $y \prec b \prec z$, and by (2) for $\prec$, $y \prec x$, in which case $a \prec^2 x$, or $x \prec z$, in which case $x \prec^2 c$. Hence (2) holds for $\prec^2$. To prove that (1) holds for $\prec^2$, suppose $a \prec^2 x$ and $b \prec^2 y$, say $a \prec c \prec x$ and $b \prec d \prec y$. According to (2) for $\prec$, either $a \prec d$ or $d \prec x$. If $a \prec d$ then $a \prec^2 y$, and if $d \prec x$ then $b \prec^2 x$. Hence (1) holds for $\prec^2$, so $\prec^2$ is a semiorder on X.

The second part of Theorem 1 follows immediately from the first part and remarks made prior to the theorem. □

Along with $\sim$ and $\approx$, several other binary relations defined on the basis of $\prec$ on X will play a prominent role in our analysis of interval orders. The two most important relations are the left and right compositions of $\sim$ and $\prec$ defined as follows:

$$\prec^- = (\sim)(\prec): x \prec^- y \quad \text{if } x \sim a \prec y \quad \text{for some } a;$$

$$\prec^+ = (\prec)(\sim): x \prec^+ y \quad \text{if } x \prec b \sim y \quad \text{for some } b.$$

We shall see later that $\prec^-$ essentially governs the placements of left ends of intervals in an interval representation of $(X, \prec)$, and that $\prec^+$ essentially governs the placements of right ends of intervals. This is illustrated by Fig. 2.2, where the lines suggest intervals in $(\mathbb{R}, <)$ or some other linearly ordered set. Here and later we label intervals with their points from X and displace some

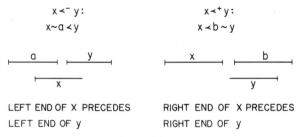

Figure 2.2

vertically for visual comprehension. A little vertical bar denotes an endpoint; no such bar leaves the endpoint unsettled. Interval x lies wholly to the left of y if $x \prec y$. Intervals x and y touch or overlap if $x \sim y$.

As with $\sim = sc(\prec)$, we define $\sim^{-} = sc(\prec^{-})$ and $\sim^{+} = sc(\prec^{+})$. The first of our two basic structural theorems for interval orders and semiorders is

Theorem 2. *Suppose $(X, \prec)$ is an interval order. Then $\prec^{-}$ and $\prec^{+}$ are weak orders on X, and $\approx = (\sim^{-} \cap \sim^{+})$. If $(X, \prec)$ is a semiorder, then $\prec^{-} \cup \prec^{+}$ is a weak order on X that equals $S_0(\prec)$, the sequel of $\prec$.*

Proof. Let $(X, \prec)$ be an interval order. Then $\prec^{+}$ is asymmetric, for otherwise $x \prec^{+} y$ and $y \prec^{+} x$ give $x \prec a \sim y$ and $y \prec b \sim x$, which contradict (1). For negative transitivity, suppose $x \prec^{+} y$, say $x \prec b \sim y$, and consider any z in X. If not($x \prec^{+} z$), then $z \prec b$ and therefore $z \prec^{+} y$. The proof for $\prec^{-}$ is similar.

If $x \sim^{-} y$ and $x \sim^{+} y$ then $x \sim a \Rightarrow \text{not}(a \prec y)$, and $a \sim x \Rightarrow \text{not}(y \prec a)$, so $x \sim a \Rightarrow y \sim a$. Conversely, if $x \not\sim^{-} y$ or $x \not\sim^{+} y$, then $x \not\approx y$. Thus $\approx = (\sim^{-} \cap \sim^{+})$.

To show that $\prec^{-} \cup \prec^{+}$ is a weak order when $(X, \prec)$ is a semiorder, consider asymmetry first. $(x \prec^{-} y, y \prec^{-} x)$ and $(x \prec^{+} y, y \prec^{+} x)$ are prohibited by the first part of the theorem. If $(x \prec^{-} y, y \prec^{+} x)$ then $(x \sim a \prec y, y \prec b \sim x)$, which violates (2). For negative transitivity, not$(x(\prec^{-} \cup \prec^{+})y)$ $\Rightarrow$ (not($x \prec^{-} y$), not($x \prec^{+} y$)), and not$(y(\prec^{-} \cup \prec^{+})z) \Rightarrow$ (not($y \prec^{-} z$), not($y \prec^{+} z$)), so negative transitivity for $\prec^{-}$ and $\prec^{+}$ gives (not($x \prec^{-} z$), not($x \prec^{+} z$)), or not$(x(\prec^{-} \cup \prec^{+})z)$.

When $(X, \prec)$ is a semiorder, the demonstrations in the preceding paragraph show that $[(\sim)(\prec) \cup (\prec)(\sim)] \cap cd[(\sim)(\prec) \cup (\prec)(\sim)] = \emptyset$, hence that $\prec^{-} \cup \prec^{+}$ equals $S_0(\prec)$ as defined in Section 1.3. □

The latter part of Theorem 2 shows that $x \prec^{-} y$ and $y \prec^{+} x$ cannot occur when $(X, \prec)$ is a semiorder. However, this can happen when $(X, \prec)$ is an interval order, and, when it does occur, it implies that the interval used to represent y is properly included in the interval used to represent x. More

Basic Structures

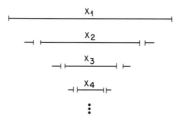

Figure 2.3 Nested intervals.

generally, we could have $x_1 \prec^- x_2 \prec^- x_3 \prec^- \cdots$ and $\cdots \prec^+ x_3 \prec^+ x_2 \prec^+ x_1$ for a nested sequence of intervals, as pictured in Fig. 2.3.

Our second basic structural theorem uses a compound binary relation formed from $\prec^-$ on X and $\prec^+$ on X. The domain of the new relation is the union of two disjoint copies of X, say

$$X^- = \{(x, -): x \in X\},$$
$$X^+ = \{(x, +): x \in X\}.$$

We write $(x, -)$ as x^- and $(x, +)$ as x^+. Our intention is to associate x^- with the left endpoint of an interval for x, and to associate x^+ with its right endpoint.

We define $\prec_0$ on $X^- \cup X^+$ in four parts as follows:

1. $x^- \prec_0 y^- \Leftrightarrow x \prec^- y$.
2. $x^+ \prec_0 y^+ \Leftrightarrow x \prec^+ y$.
3. $x^+ \prec_0 y^- \Leftrightarrow x \prec y$.
4. $x^- \prec_0 y^+ \Leftrightarrow x \prec y$ or $x \sim y$.

Line-for-line: **1**, the left end of x precedes the left end of y if $x \prec^- y$; **2**, the right end of x precedes the right end of y if $x \prec^+ y$; **3**, the right end of x precedes the left end of y if $x \prec y$; **4**, the left end of x precedes the right end of y if not($y \prec x$). Part 4 could be modified to allow the intervals for x and y to barely touch when $x \sim y$, but to do so would introduce complications that are best avoided at this time.

The following theorem takes $\sim_0 = sc(\prec_0)$ and adopts the notation for equivalence classes (justified by Theorems 1.2 and 2.2) given by

$$X^-/\sim^- = \{A^-: A \in X/\sim^-\},$$
$$X^+/\sim^+ = \{A^+: A \in X/\sim^+\}.$$

Theorem 3. *Suppose $(X, \prec)$ is an interval order. Then $\prec_0$ is a weak order on $X^- \cup X^+$;*

$$(X^- \cup X^+)/\sim_0 = (X^-/\sim^-) \cup (X^+/\sim^+);$$

$A^- \in X^-/\sim^- \Rightarrow A^- \prec_0 B^+$ for some $B^+ \in X^+/\sim^+$; $A^+ \in X^+/\sim^+ \Rightarrow B^- \prec_0 A^+$ for some $B^- \in X^-/\sim^-$; $A^- \prec_0 B^-$ for $A^-, B^- \in X^-/\sim^- \Rightarrow A^- \prec_0 C^+ \prec_0 B^-$ for some $C^+ \in X^+/\sim^+$; and $A^+ \prec_0 B^+$ for $A^+, B^+ \in X^+/\sim^+ \Rightarrow A^+ \prec_0 C^- \prec_0 B^+$ for some $C^- \in X^-/\sim^-$.

Proof. Let $(X, \prec)$ be an interval order. We prove first that $\prec_0$ is a weak order on $X^- \cup X^+ = \{a, b, c, \ldots\}$.

Asymmetry. Assume $a \prec_0 b$. If $a, b \in X^-$ or $a, b \in X^+$, then not($b \prec_0 a$) by Theorem 2. If $(a, b) = (x^-, y^+)$, then not($y \prec x$) by part 4, and thus not($y^+ \prec_0 x^-$) by part 3. If $(a, b) = (x^+, y^-)$, then $x \prec y$ by part 3, hence not($y^- \prec_0 x^+$) by part 4.

Negative transitivity. Let $\precsim = \prec \cup \sim$. We suppose that negative transitivity fails with

$$a \prec_0 c, \quad \text{not}(a \prec_0 b), \quad \text{not}(b \prec_0 c),$$

and obtain a contradiction. If $a, b, c \in X^-$ or $a, b, c \in X^+$, the contradiction is assured by Theorem 2. The other six cases follow.

1. $(a, b, c) = (x^-, y^-, z^+)$. Then $x \precsim z$, not($x \prec^- y$), $z \prec y$. However, $x \precsim z \prec y \Rightarrow x \prec^- y$, a contradiction.
2. $(a, b, c) = (x^-, y^+, z^-)$. Then $x \prec^- z, y \prec x, z \precsim y$. By $x \prec^- z, x \sim w \prec z$ for some w. But this and $z \precsim y \prec x$ contradict (1).
3. $(a, b, c) = (x^+, y^-, z^-)$. Similar to case 1.
4. $(a, b, c) = (x^-, y^+, z^+)$. Then $x \precsim z, y \prec x$, not($z \prec^+ z$). However, $y \prec x \precsim z \Rightarrow y \prec^+ z$, a contradiction.
5. $(a, b, c) = (x^+, y^-, z^+)$. Then $x \prec^+ z, y \precsim x, z \prec y$. By $x \prec^+ z, x \prec w \sim z$ for some w. This and $z \prec y \precsim x$ contradict (1).
6. $(a, b, c) = (x^+, y^+, z^-)$. Similar to case 4.

This completes the proof that $\prec_0$ is a weak order. By parts 3 and 4, $a \in X^-$ and $b \in X^+$ imply $a \not\sim_0 b$. Hence, by parts 1 and 2, $a \sim_0 b$ if and only if either $\{a, b\} = \{x^-, y^-\}$ and $x \sim^- y$, or $\{a, b\} = \{x^+, y^+\}$ and $x \sim^+ y$. Therefore the equivalence classes in $(X^- \cup X^+)/\sim_0$ (see Theorem 1.2) are the classes in $X^-/\sim^-$ plus the classes in $X^+/\sim^+$.

Since $x^- \prec_0 x^+$ by part 4, it follows that for every A^- in $X^-/\sim^-$ there is a B^+ in $X^+/\sim^+$ for which $A^- \prec_0 B^+$, and for every B^+ there is an A^- such that $A^- \prec_0 B^+$.

Finally, suppose $A^- \prec_0 B^-$ with $x^- \in A^-$ and $y^- \in B^-$, so $x^- \prec_0 y^-$ and $x \prec^- y$. Then $x \sim z \prec y$ for some z, so $x^- \prec_0 z^+ \prec_0 y^-$ by parts 4 and 3, and $A^- \prec_0 C^+ \prec_0 B^-$ when $z^+ \in C^+$. Similarly, $A^+ \prec_0 B^+$ implies $A^+ \prec_0 C^- \prec_0 B^+$ for some C^-. □

We shall refer to $\prec_0$ on $X^- \cup X^+$ as the *conjoint weak order* of interval order $(X, \prec)$. It will be used in Section 2.4 to lead off our representation theorems for interval orders and semiorders. Prior to that, we consider a few more structural properties. The first of these concerns semiorders.

Theorem 4. *Suppose $(X, \prec)$ is a semiorder. Then there is a linear order $\prec_0^*$ on $X^- \cup X^+$ such that $\prec_0 \subseteq \prec_0^*$ and, for all $x, y \in X$,*

$$x^- \prec_0^* y^- \Leftrightarrow x^+ \prec_0^* y^+.$$

Proof. Let $(X, \prec)$ be a semiorder with conjoint weak order $\prec_0$ on $X^- \cup X^+$.

Suppose first that $\approx$ is empty apart from $x \approx x$, so that it is never true for distinct x and y in X that $x \approx y$, or $(x^- \sim^- y^-, x^+ \sim^+ y^+)$ or $(x^- \sim_0 y^-, x^+ \sim_0 y^+)$ in view of Theorems 2 and 3. Define $\prec_0^*$ by

$$a \prec_0^* b \quad \text{if } a \prec_0 b;$$

$$x^- \prec_0^* y^- \quad \text{if } x^- \sim_0 y^- \text{ and } x^+ \prec_0 y^+;$$

$$x^+ \prec_0^* y^+ \quad \text{if } x^+ \sim_0 y^+ \text{ and } x^- \prec_0 y^-.$$

What this does is to linearly order every equivalence class in $X^-/\sim^-$ and in $X^+/\sim^+$ in such a way that $x^- \prec_0^* y^- \Leftrightarrow x^+ \prec_0^* y^+$. It is easily seen that $\prec_0^*$ is a linear order on $X^- \cup X^+$ and that the conclusion of Theorem 4 holds if either $x^- \sim_0 y^-$ or $x^+ \sim_0 y^+$. Suppose the conclusion fails, with $x^- \prec_0^* y^-$ and $y^+ \prec_0^* x^+$. Then $x^- \prec_0 y^-$ and $y^+ \prec_0 x^+$, so that $x \sim a \prec y$ and $y \prec b \sim x$ for some $a, b \in X$ by parts 1 and 2 in the definition of $\prec_0$. But then $a \prec y \prec b$ with $x \sim a$ and $x \sim b$, which violates (2). Hence the theorem holds in the present case.

Suppose next that $x \approx y$ for some distinct x and y. Then the proof in the preceding paragraph applies with no essential changes to $X/\approx$, that is, to the equivalence classes in X determined by $\approx$. In view of $x \approx y \Leftrightarrow (x^- \sim_0 y^-, x^+ \sim_0 y^+)$, the x^- and x^+ in each A^- and A^+ that corresponds to an $A \in X/\approx$ can be linearly ordered in the same way for each such A to provide a linear order on all of $X^- \cup X^+$ that satisfies the theorem. □

2.3 MAGNITUDES AND CHARACTERISTIC MATRICES

Theorem 4 shows that, when $(X, \prec)$ is a semiorder, its conjoint weak order $\prec_0$ can be extended to a linear order $\prec_0^*$ on $X^- \cup X^+$ with the very special property that the restriction of $\prec_0^*$ to X^- is isomorphic in the natural way to its restriction to X^+.

A similar isomorphism may fail to hold for the equivalence classes in $X^-/\sim^-$ and $X^+/\sim^+$ ordered by $\prec_0$. That is, $(X^-/\sim^-, \prec_0) \cong (X^+/\sim^+, \prec_0)$ can be false, even when $(X, \prec)$ is a semiorder. An obvious example for interval orders takes X as the set of all nondegenerate intervals on $\mathbb{R}$ that have rational left endpoints and irrational right endpoints, with $x \prec y$ if $\sup x \prec \inf y$. Then $(X^-/\sim^-, \prec_0)$ is isomorphic to the rational numbers under their natural order $\prec$, and $(X^+/\sim^+, \prec_0)$ is isomorphic to the irrational numbers under their natural order. Since the rationals are countable and the irrationals are uncountable, $(X^-/\sim^-, \prec_0) \not\cong (X^+/\sim^+, \prec_0)$.

An example for a semiorder is illustrated by the interval diagram of Fig. 2.4, which indicates a denumerable number of intervals. We take $x \prec y$ if the interval for x completely precedes the interval for y. Since $x \prec y \prec z$ never occurs, (2) holds trivially. Since $(X^-/\sim^-, \prec_0)$ has a first member but $(X^+/\sim^+, \prec_0)$ has no first member, $(X^-/\sim^-, \prec_0) \not\cong (X^+/\sim^+, \prec_0)$.

The following easy corollary (proof left to the reader) of Theorem 3 shows that such failures of isomorphism can occur only when $(X^- \cup X^+)/\sim_0$ is infinite.

Theorem 5. *Suppose $(X, \prec)$ is an interval order and $(X^- \cup X^+)/\sim_0$ is finite. Then $X^-/\sim^-$ and $X^+/\sim^+$ have the same cardinality, say m, and if $(X^-/\sim^-, \prec_0) = \{A_1^- \prec_0 A_2^- \prec_0 \cdots \prec_0 A_m^-\}$ and $\{X^+/\sim^+, \prec_0\} = \{B_1^+ \prec_0 B_2^+ \prec_0 \cdots \prec_0 B_m^+\}$, then*

$$A_1^- \prec_0 B_1^+ \prec_0 A_2^- \prec_0 B_2^+ \prec_0 \cdots \prec_0 A_m^- \prec_0 B_m^+.$$

Given the hypotheses of Theorem 5, we shall refer to $|X^-/\sim^-|$ as the *magnitude* of the interval order. When $(X, \prec)$ is a magnitude-m interval order, it follows from Theorem 5 and the definition of $\prec_0$ that there are partitions $\{A_1, \ldots, A_m\}$ and $\{B_1, \ldots, B_m\}$ of X such that

$$(X/\sim^-, \prec^-) = \{A_1 \prec^- A_2 \prec^- \cdots \prec^- A_m\},$$
$$(X/\sim^+, \prec^+) = \{B_1 \prec^+ B_2 \prec^+ \cdots \prec^+ B_m\}.$$

Given these ordered partitions, we define the *characteristic matrix* **M** =

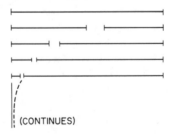

Figure 2.4 Denumerable semiorder.

Magnitudes and Characteristic Matrices

$M(X, \prec)$ of the interval order as the upper triangular 0-1 matrix defined on $\{(i, j): 1 \leq i \leq j \leq m\}$ by

$$M_{ij} = 1 \quad \text{if } x \in A_i \cap B_j \text{ for some } x \in X;$$

$$M_{ij} = 0 \quad \text{otherwise.}$$

Thus $M_{ij} = 1$ if some point in X is in both the ith left endpoint class and the jth right endpoint class.

Figure 2.5 pictures a magnitude-7 interval order and its characteristic matrix. The 1's in the matrix are subscripted by their corresponding points from X. Casual inspection of the matrix shows that there is at least one 1 in every row and in every column. This of course is no accident; it is an immediate consequence of Theorem 5, the preceding paragraph, and the definition of $\prec_0$.

Moreover, an upper-triangular $m \times m$ 0-1 matrix is the characteristic matrix of an interval order *if* and only if it has at least one 1 in every row and column, and two interval orders without $\approx$ holding between distinct points are isomorphic if and only if they have the same characteristic matrix (given finite magnitudes). In addition, a characteristic matrix for a chain has 1's along the diagonal and 0's elsewhere, and a characteristic matrix corresponds to a

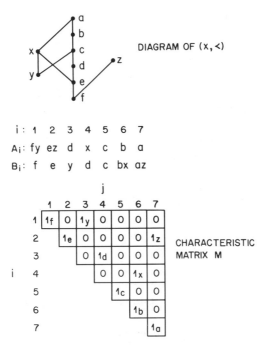

Figure 2.5 A magnitude-7 interval order.

semiorder if and only if no 1 is above and to the right of another 1. Proofs of these simple but useful facts are left to the reader. We shall return to them in later chapters.

We note also that the maximal antichains in a finite interval order follow from the structure described by Theorem 5. In particular, with

$$A_1^- \prec_0 B_1^+ \prec_0 \cdots \prec_0 A_m^- \prec_0 B_m^+,$$

the set of maximal antichains is

$$\left\{ \bigcup_{i=1}^{j} A_i \setminus \bigcup_{i=1}^{j-1} B_i : j = 1, \ldots, m \right\},$$

which is the same as

$$\left\{ \bigcup_{i=j}^{m} B_i \setminus \bigcup_{i=j+1}^{m} A_i : j = 1, \ldots, m \right\}$$

since

$$\bigcup_{i=1}^{j} A_i \setminus \bigcup_{i=1}^{j-1} B_i = \bigcup_{i=j}^{m} B_i \setminus \bigcup_{i=j+1}^{m} A_i.$$

For example, A_1 is a maximal antichain; when the $x \in B_1$ are dropped from A_1, we get a nonmaximal antichain, but when A_2 is added to $A_1 \setminus B_1$, a second maximal antichain appears; when B_2 is subtracted from $(A_1 \cup A_2) \setminus B_1$ and then A_3 is added, a third maximal antichain appears; and so on through A_m.

In terms of the characteristic matrix **M**, each maximal antichain consists of the points that map into rows 1 through j, minus the points that map into columns 1 through $j - 1$. For example, the maximal antichains for Fig. 2.5 are $\{f, y\}$, $\{y, e, z\}$, $\{y, z, d\}$, $\{z, d, x\}$, $\{z, x, c\}$, $\{z, x, b\}$, and $\{z, a\}$. These are of course the points in the maximal rectangular submatrices of **M**.

As a consequence of the preceding discussion, we have

Corollary 1. *The magnitude of a finite interval order equals the number of its maximal antichains.*

2.4 REPRESENTATION THEOREMS

We begin with general theorems for representations by intervals of linearly ordered sets, and then consider real interval representations.

Theorem 6. $(X, \prec)$ *is an interval order if and only if there is a linearly ordered set* $(Y, <_0)$ *and a mapping F from X into closed intervals in Y such that, for all*

$x, y \in X$,
$$x \prec y \Leftrightarrow F(x) <_0 F(y).$$

Proof. If the representation holds then it is easily seen that $\prec$ is irreflexive and satisfies (1). Conversely, if $(X, \prec)$ is an interval order, let $Y = (X^- \cup X^+)/\sim_0$, ordered by the conjoint weak order $\prec_0$. By Theorems 3 and 1.2, $(Y, \prec_0)$ is a linearly ordered set. Let $[x^-]$ denote the equivalence class in Y that contains x^- (similarly for $[x^+]$), and define
$$F(x) = \{a \in Y: [x^-] \lesssim_0 a \lesssim_0 [x^+]\}.$$
Then, by Theorem 3 and the definition of $\prec_0$ we have $x \prec y \Leftrightarrow x^+ \prec_0 y^- \Leftrightarrow F(x) \prec_0 F(y)$. □

Theorem 7. *$(X, \prec)$ is a semiorder if and only if there is a linearly ordered set $(Y, <_0)$ and a mapping F from X into closed intervals in Y such that, for all $x, y \in X$,*
$$x \prec y \Leftrightarrow F(x) <_0 F(y),$$
$$\inf F(x) <_0 \inf F(y) \Leftrightarrow \sup F(x) <_0 \sup F(y).$$

Proof. The representation is easily seen to imply (2). Conversely, if $(X, \prec)$ is a semiorder, let $Y = X^- \cup X^+$, let $\prec_0^*$ be a linear order on Y with the properties guaranteed by Theorem 4, and define $F(x)$ as $\{a \in Y: x^- \lesssim_0^* a \lesssim_0^* x^+\}$. Then $x \prec y \Leftrightarrow x^+ \prec_0 y^- \Leftrightarrow x^+ \prec_0^* y^- \Leftrightarrow F(x) \prec_0^* F(y)$ and, by the final part of Theorem 4, $\inf F(x) \prec_0^* \inf F(y) \Leftrightarrow \sup F(x) \prec_0^* \sup F(y)$. □

Since a partial semiorder need not be an interval order, it might not have an interval representation in the sense of Theorem 6. However, Theorems 1 and 7 show that the $\prec^2$ part of a partial semiorder can be represented in the manner of Theorem 7. It may be possible to augment the $\prec^2$ representation by an auxiliary representation for the diagram $\prec \setminus \prec^2$ that uses the same F, but I shall not pursue this here.

We now turn to real representations.

Theorem 8. *If $(X, \prec)$ is an interval order and $X/\approx$ is countable, then the conclusion of Theorem 6 holds with $(Y, <_0) = (\mathbb{R}, <)$. If $(X, \prec)$ is a semiorder and $X/\approx$ is countable, then the conclusion of Theorem 7 holds with $(Y, <_0) = (\mathbb{R}, <)$.*

Proof. If we first reduce X in the semiorder case by dividing out $\approx$, then the linearly ordered Y's in the proofs of Theorems 6 and 7 are countable and by Theorem 1.4 are isomorphic to subsets of $(\mathbb{R}, <)$. The desired conclusions follow by mapping the closed Y intervals into their corresponding real intervals. □

Extensions of Theorem 8 for cases in which $X/\approx$ is not countable will be discussed in Chapter 7. At this point we note one other real interval theorem, but will not prove it until Chapter 7. It was first proved by Scott and Suppes (1958) and has since been proved in several different ways.

Theorem 9 (Scott and Suppes). *If $(X, \prec)$ is a finite semiorder, then there is a real valued function f on X such that, for all $x, y \in X$,*

$$x \prec y \Leftrightarrow f(x) + 1 < f(y).$$

In other words, if X is finite, then the real interval representation for semiorders of Theorem 8 can be further specialized so that all intervals $F(x)$ have unit length. There is nothing special about the use of 1 in the theorem since any other fixed positive length could be used.

The fixed-length representation of Theorem 9 is sometimes valid for infinite semiorders. This is true, for example, for the denumerable semiorder pictured in Fig. 2.4. However, other countable semiorders cannot be represented by real intervals of a single length. A simple example is $X = \{[n, n+1]: n \in \mathbb{Z}\} \cup \{\omega\}$, with $\prec = <$ on the intervals in X and $x \prec \omega$ for all x in $X \setminus \{\omega\}$.

In later chapters we shall refer to a mapping F that satisfies Theorem 6 as a *representation* of $(X, \prec)$. When $(Y, <_0) = (\mathbb{R}, <)$, as in Theorem 8, the interval $F(x)$ will sometimes be written as

$$F(x) = [f(x), f(x) + \rho(x)]$$

where $f: X \to \mathbb{R}$ is the *location function* and $\rho: X \to \mathbb{R}$ is the *length function* of the representation. The real form will generally be used for finite X, and ρ will often be assumed to be strictly positive. In all cases, $\rho \geq 0$, and a pair (f, ρ) of real functions that satisfies

$$x \prec y \Leftrightarrow f(x) + \rho(x) < f(y), \quad \text{for all } x, y \in X,$$

will also be referred to as a representation.

2.5 WEAK ORDER EXTENSIONS

In this section we assume that X is finite and that $(X, \prec)$ is a poset or an interval order. The purpose of the section is to comment on two ways to extend $(X, \prec)$ to a weakly ordered set $(X, \prec')$ with $\prec \subseteq \prec'$. The motivation for such

Weak Order Extensions

extensions is provided by situations in which X is believed to be weakly or linearly ordered but the discriminatory power of the measurement process that seeks to uncover this order is limited. The measurement data $\prec$ give only a partial picture of the underlying order, and we seek an extension process that is likely to correctly identify ordered pairs of that order which are not part of the data. Our concern with weak order extensions reflects the fact that if $x \approx y$ then all reasonable methods for extending $\prec$ will have x equivalent to y in the extension since there is nothing in the data that distinguishes between them.

One intuitively sensible method for extending $\prec$ is based on the sequel operation S_0 defined in Section 1.3. Recall that if $x \sim y$ and $x \neq y$, then (x, y) is in $S_0(\prec)$ if and only if either $x \sim a \prec y$ or $x \prec a \sim y$ for some $a \in X$, and it is not true that $y \sim b \prec x$ or $y \prec b \sim x$ for any $b \in X$. Moreover, according to Theorems 1.1 and 1.2, $S_0(\prec)$ is a partial order that includes $\prec$, and it properly includes $\prec$ if and only if $\prec$ is not a weak order. Hence, by applying S_0 iteratively, we eventually obtain a weak order that includes $\prec$. We shall denote this weak order as $S_*(\prec)$, and will focus on it until later in the section.

To be precise, let

$$S_i(\prec) = S_0(S_{i-1}(\prec)) \quad \text{for } i = 1, 2, \ldots,$$

and let $S_*(\prec) = \lim S_i(\prec)$. If $\prec$ is a weak order then $S_*(\prec) = \prec$ by Theorem 1.2, and in this case we say that $(X, \prec)$ has *sequel degree* 0. Otherwise, if $(X, \prec)$ is not a weakly ordered set, the *sequel degree* η of $(X, \prec)$, which we also write as $\eta(X, \prec)$ or as $\eta(\prec)$, is the smallest $k \in \{1, 2, \ldots\}$ for which $S_{k-1}(\prec) = S_k(\prec)$.

Theorem 2 shows that $\eta \leq 1$ when $(X, \prec)$ is a semiorder. Figure 2.6 describes the successive $S_i(\prec)$ for a poset that has $\eta = 3$. There is no upper bound on η for finite posets, and in fact no upper bound on η for finite interval orders. This is shown most easily by $(X, \prec) = \{a\} \cup \{x_1 \prec x_2 \prec \cdots \prec x_m\}$, where $\prec$ linearly orders $X \setminus \{a\}$ and $a \sim x_i$ for all i. $S_0(\prec)$ has $x_1 S_0(\prec) a S_0(\prec) x_m$, but no other new pairs are added by $S_0(\prec)$; $S_1(\prec)$ adds $x_2 S_1(\prec) a S_1(\prec) x_{m-1}$; and so forth. Thus $\eta(\prec) = (m-1)/2$ if m is odd, and $\eta(\prec) = m/2$ if m is even.

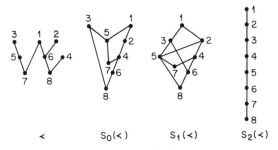

Figure 2.6 Sequels of a poset.

Another example for posets, which relates to many other standard examples in the literature, takes $(X, \prec)$ equal to $(2^{\{1,\ldots,n\}}, \subset)$, the lattice of all subsets of $\{1,\ldots,n\}$ partially ordered by proper inclusion.

Theorem 10. $\eta(2^{\{1,\ldots,n\}}, \subset) = n - 2$ for $n \geq 2$, and $S_*(\subset) = \{(A, B): A, B \subseteq \{1,\ldots,n\}$ and $|A| < |B|\}$.

Proof. Let $(X, \prec) = (2^{\{1,\ldots,n\}}, \subset)$ with $n \geq 2$, and let $A_j = \{A \subseteq \{1,\ldots,n\}: |A| = j\}$. Every maximal chain in $(X, \prec)$ has the form $\{\emptyset \prec x_1 \prec \cdots \prec x_{n-1} \prec \{1,\ldots,n\}\}$ with $x_j \in A_j$ for all j. In addition, $A_j \sim A_j$ for all j. The symmetry of the lattice structure then forces $A_0 S_*(\prec) A_1 S_*(\prec) \cdots S_*(\prec) A_n$.

To verify $\eta(X, \prec) = n - 2$, note first that $\eta = 0$ for $n = 2$ since $A_0 \prec A_1 \prec A_2$ in this case. Assume henceforth that $n \geq 3$, and consider $x \in A_j$ and $y \in A_k$ with $1 \leq j < k \leq n - 1$, $x \not\subseteq y$, and $a \in x \setminus y$. Then clearly $x(\sim)(\prec)y$. If $(j, k) = (1, n - 1)$ then not$[y(\sim)(\prec)x$ or $y(\prec)(\sim)x]$, but if $k - j \leq n - 3$ then either $j = 1$ and $y \prec y \cup \{b\} \sim x$ with $b \notin x \cup y$, or $j \geq 2$ and $y \sim \{a\} \prec x$. Therefore

$$S_0(\prec) = \prec \cup (A_1 \times A_{n-1}).$$

Thus, if $n = 3$, then $S_1(\prec) = S_0(\prec) = \{A_0 \prec A_1 S_0(\prec) A_2 \prec A_3\}$ and $\eta = n - 2 = 1$.

Assume henceforth that $n > 3$, and let

$$R(\alpha) = \cup \{A_j \times A_k: 1 \leq j < k \leq n - 1 \text{ and } k - j = \alpha\}.$$

We have just proved that $S_0(\prec) = \prec \cup R(n - 2)$. We claim that, for $i = 1, \ldots, n - 3$,

$$S_i(\prec) = S_{i-1}(\prec) \cup R(n - (i + 2)).$$

Suppose this is true for all i less than a fixed h in $\{1,\ldots,n-3\}$, with $S_{-1}(\prec) = \prec$. We show that it is true also at h. By the induction hypothesis,

$$S_{h-1}(\prec) = \prec \cup [R(n - 2) \cup R(n - 3) \cup \cdots \cup R(n - h - 1)].$$

Let I denote the symmetric complement of $S_{h-1}(\prec)$, and let $P = S_{h-1}(\prec)$. Consider $x \in A_j$ and $y \in A_k$ with $1 \leq j < k \leq n - 1$, $k - j \leq n - h - 2$, and $x \not\subseteq y$. Clearly $xIPy$. If $k - j = n - h - 2$, then not$(yIPx$ or $yPIx)$, and therefore $R(n - h - 2)$ is in $S_h(\prec)$. On the other hand, if $k - j < n - h - 2$, then either $j = 1$, in which case $yP(y \cup \{b\})Ix$ with $b \notin x \cup y$, or $j \geq 2$, in which case $yIzPx$ for any $z \in A_{j-1}$ for which $z \subset x$ and $z \not\subseteq y$. Hence no new (x, y) with $|y| - |x| < n - h - 2$ are added to $S_{h-1}(\prec)$ to get $S_h(\prec)$, and therefore

$$S_h(\prec) = S_{h-1}(\prec) \cup R(n - h - 2).$$

It follows that $S_{n-4}(\prec) \subset S_{n-3}(\prec) = S_{n-4}(\prec) \cup R(1)$. Moreover, $S_{n-3}(\prec)$ is clearly the weak order $S_*(\prec)$, and $S_{n-4}(\prec) \subset S_{n-3}(\prec) = S_{n-2}(\prec) = S_*(\prec)$ therefore yields $\eta(X, \prec) = n - 2$. □

A second method for extending $\prec$ to a weak order is a simple one-step procedure based on a global score function

$$g(x) = |\{z: z \prec x\}| - |\{z: x \prec z\}|,$$

so that $g(x)$ is the number of points below x in $(X, \prec)$ minus the number above x in $(X, \prec)$. If $x \prec y$ then clearly $g(x) < g(y)$. Thus, when $G_*(\prec)$ is defined by

$$xG_*(\prec)y \quad \text{if } g(x) < g(y),$$

it follows that $G_*(\prec)$ is a weak order that includes $\prec$. Our next theorem (proof left to the reader) shows how G_* relates to the sequel operation.

Theorem 11. *Suppose $(X, \prec)$ is a poset. Then, for all $x, y \in X$,*

$$xG_*(\prec)y \Leftrightarrow |\{z: x \prec z \sim y \text{ or } x \sim z \prec y\}| < |\{z: y \prec z \sim x \text{ or } y \sim z \prec x\}|$$

and $S_0(\prec) \subseteq G_*(\prec)$. If $(X, \prec)$ is a semiorder, then $G_*(\prec) = S_*(\prec)$.

Thus, the two methods are the same for semiorders as well as for some other cases, including the lattice poset of Theorem 10. For Fig. 2.6, $G_*(\prec)$ is the same as $S_2(\prec)$ except that $\{2,3\}$ is an equivalence class for G_*.

Fishburn and Gehrlein (1974, 1975) report simulation studies that compare the abilities of $S_*(\prec)$ and $G_*(\prec)$ to correctly identify pairs in an underlying linear order that are not already in $\prec$. Posets and interval orders were randomly chosen from a linearly ordered set on 10, 15, or 20 points, then $S_*(\prec)$ and $G_*(\prec)$ were determined for each chosen $\prec$. An error score was computed for each case by the formula

$$|\{(x, y): xLy \text{ and } yWx\}| + \tfrac{1}{2}|\{(x, y): xLy \text{ and } xsc(W)y\}|,$$

where L is the linear order used to generate the data $\prec$, and W is a weak order constructed from $\prec$ by one method or the other. Our error score is tantamount to the measure of distance between weak orders (L and W) that was axiomatized by Kemeny (1959) and Kemeny and Snell (1962) and was subsequently generalized to posets by Bogart (1973a). Several different procedures were used to generate the posets and interval orders from L, and runs for each case ($|X|$, generating procedure) consisted of 500 or 1000 or more trials.

When the generating procedure was designed to yield a partial order but not necessarily an interval order, G_* generally had a lower average error score than S_*. On the other hand, when the procedure was explicitly designed to produce an interval order, the average error score for S_* was less than that for G_* when the data were relatively sparse, that is, when the number of ordered pairs in $\prec$ was less than about $(0.6)\binom{|X|}{2}$. However, when the procedure to generate interval orders gave $|\prec|$ greater than about $(0.6)\binom{|X|}{2}$, the earlier finding of lower average error score for G_* was observed.

3

Interval Graphs

This chapter defines interval graphs and indifference graphs in terms of nonempty intersections of intervals of linearly ordered sets and notes their intimate ties to interval orders and semiorders. Equivalent characterizations of these graphs in terms of comparability graphs and forbidden subgraphs are then established. The last section of the chapter includes two characterizations of interval graphs that have unique agreeing interval orders up to duality of the ordering relation.

3.1 DEFINITIONS AND EXAMPLES

The following definitions adhere to the interval notions used in Section 2.4. Reflexive graphs are employed because of the natural correspondences described later in Theorem 1.

An *interval graph* is a reflexive graph $(X, \sim)$ for which there exists a mapping F from X into closed intervals of a linearly ordered set $(Y, <_0)$ such that, for all $x, y \in X$,

$$x \sim y \Leftrightarrow F(x) \cap F(y) \neq \emptyset.$$

An *indifference graph* is a reflective graph $(X, \sim)$ for which there exists a mapping F from X into closed intervals of a linearly ordered set $(Y, <_0)$ such that, for all $x, y \in X$,

$$x \sim y \Leftrightarrow F(x) \cap F(y) \neq \emptyset,$$

$$\inf F(x) <_0 \inf F(y) \Leftrightarrow \sup F(x) <_0 \sup F(y).$$

Figure 3.1 pictures two interval graphs and three graphs that are not interval graphs. Unless noted otherwise, all drawn graphs are assumed to be reflexive, but loops will not be shown. Graph I is an indifference graph since it has an interval representation, shown to the right of the graph, that satisfies the defining conditions for indifference graphs. Graph II (the bipartite graph $K_{1,3}$) is the smallest interval graph that is not also an indifference graph. Every other four-point interval graph that is not an indifference graph is isomorphic to graph II.

The partial interval representations for the other three graphs in the figure show why they are not interval graphs. Graph IIIa, the simple cycle of length 4 (or the complete bipartite graph $K_{2,2}$), is the only four-point graph that is not an interval graph. For IIIc, note that once the intervals for x, a, b, and y have been drawn, c's interval must fit strictly between x and y. But then it is impossible to position z so that it intersects only c.

Graph IIIc is an example of a graph $(X, \sim)$ that has an *asteriodal triple* (Lekkerkerker and Boland, 1962). This is a three-point independent set $\{x, y, z\}$ such that, for each two distinct points in $\{x, y, z\}$, there is a path between the two that contains no point adjacent to ($\sim$) the third point in $\{x, y, z\}$. It is easily seen that a graph with an asteroidal triple cannot be an interval graph.

Note also that the *complement* of graph IIIc is not a comparability graph (Theorem 1.7) since it has an odd-length cycle

$$y\,c\,y\,a\,y\,x\,b\,x\,c\,x\,z\,a\,z\,b\,z\,y \quad \text{(length 15)}$$

with no triangular chord.

The language of representations of interval orders noted at the end of Section 2.4 extends in the natural way to interval graphs. We shall refer to a mapping F that satisfies the definition of an interval graph as a *representation*

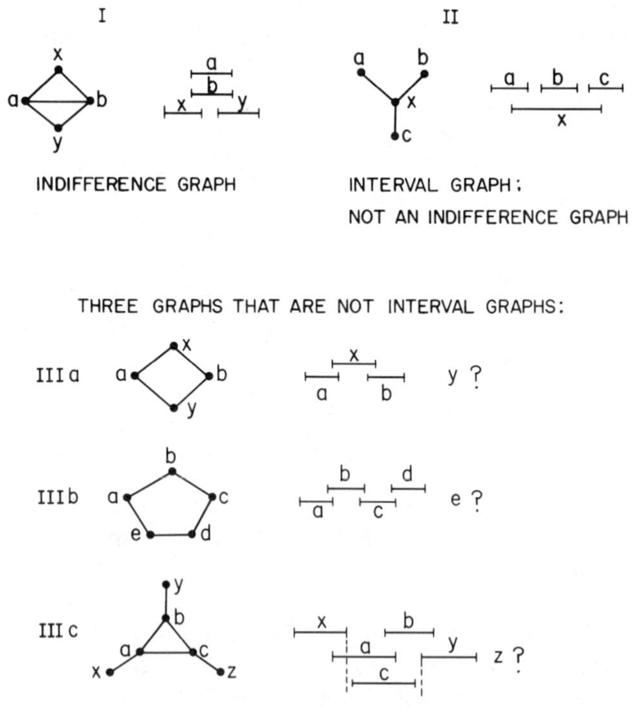

Figure 3.1 Graphs.

of the graph. This applies also to indifference graphs without necessarily assuming that F satisfies the condition that $\inf F(x) <_0 \inf F(y) \Leftrightarrow \sup F(x) <_0 \sup F(y)$, for all $x, y \in X$. When this condition does in fact hold for a particular representation F of an indifference graph, we refer to F as a *uniform representation*. Real representations (f, ρ) with

$$x \sim y \Leftrightarrow [f(x) \leq f(y) + \rho(y) \text{ and } f(y) \leq f(x) + \rho(x)],$$

for all $x, y \in X$, will often be used for finite X. As with interval orders, $\rho \geq 0$ is the length function of the representation.

If we think of a representation F of an interval graph $(X, \sim)$ as $(\{F(x): x \in X\}, \sim')$, where $F(x) \sim' F(y) \Leftrightarrow F(x) \cap F(y) \neq \emptyset$, then the representation itself is an interval graph. As such, it is a special type of *intersection graph*, which is a graph $(V, \sim')$ in which V is a family of nonempty sets and, for all $u, v \in V, u \sim' v \Leftrightarrow u \cap v \neq \emptyset$. The key feature of interval graphs thus represented is that they are intersection graphs whose elements are intervals of a linearly ordered set.

Examples of interval graphs, or of intersection graphs, whose elements are intervals of linearly ordered sets, are plentiful. In particular, the symmetric complements of the interval orders mentioned in the latter part of Section 2.1, for temporal events, inexact measurement, subjective probability, preference comparisons, and psychophysical judgment, are interval graphs. The term *indifference graph*, which was first used in the present sense by Roberts (1969), was named after the widespread use in economics of "indifference" for the symmetric complement of strict preference.

In other contexts, $\sim$ is referred to as a matching or similarity relation (Harary, 1964), or by some other convenient name. Specific reference to interval graphs in archaeological seriation is made by Kendall (1969)–see Roberts (1979, p. 255) for other references, they have been discussed by Benzer (1959) and Fulkerson and Gross (1965) in attempting to discover underlying linear arrangements in the fine structure of genes through overlaps caused by mutations, and Cohen (1978) argues that food webs of organisms in an ecological environment often approximate interval graphs. In the last case, $x \sim y$ could mean that the diets of organism types x and y overlap. Other applications, including scheduling and storage problems, are discussed by Golumbic (1980).

3.2 BASIC TIES TO INTERVAL ORDERS

The definitions of interval graphs and indifference graphs suggest close ties to interval orders and semiorders through Theorems 2.6 and 2.7. These ties are made explicit in the following theorems, the second of which is due to Roberts (1969, 1970).

Theorem 1. $(X, \sim)$ *is an interval graph if and only if there is an interval order* $(X, \prec)$ *with* $\sim = sc(\prec)$. $(X, \prec)$ *is an interval order if and only if* $\prec$ *is transitive and* $(X, sc(\prec))$ *is an interval graph.*

Theorem 2 (Roberts). $(X, \sim)$ *is an indifference graph if and only if there is a semiorder* $(X, \prec)$ *with* $\sim = sc(\prec)$. $(X, \prec)$ *is a semiorder if and only if* $\prec$ *is transitive and* $(X, sc(\prec))$ *is an indifference graph.*

Proof. Since the theorems follow easily from the definitions and Theorems 2.6 (interval orders) and 2.7 (semiorders), I shall note only the proof of Theorem 1, leaving the proof of Theorem 2 to the reader.

Let $(X, \sim)$ be an interval graph with representation F. Define $\prec$ by $x \prec y$ if $F(x) <_0 F(y)$. Then $\sim = sc(\prec)$ and, by Theorem 2.6, $(X, \prec)$ is an interval order. Conversely, if $(X, \prec)$ is an interval order with $\sim = sc(\prec)$, then the representation of Theorem 2.6 gives

$$x \sim y \Leftrightarrow \text{not}\big[F(x) <_0 F(y) \text{ or } F(y) <_0 F(x)\big] \Leftrightarrow F(x) \cap F(y) \neq \emptyset,$$

so $(X, \sim)$ is an interval graph.

For the second part of Theorem 1, it is clear that if $(X, \prec)$ is an interval order then $\prec$ is transitive and $(X, sc(\prec))$ is an interval graph. Conversely, suppose $\prec$ is transitive and $(X, sc(\prec))$ is an interval graph. The latter assumption implies that $\prec$ is irreflexive. Let $\sim = sc(\prec)$ and let F be a representation of $(X, \sim)$. If (2.1) of Section 2.1 fails, say with

$$a \prec x, \quad b \prec y, \quad \text{not}(a \prec y), \quad \text{not}(b \prec x),$$

then transitivity of $\prec$ implies $\text{not}(y \prec a)$ and $\text{not}(x \prec b)$, so $a \sim y$ and $b \sim x$. But then

$$F(a) \cap F(x) = \emptyset = F(b) \cap F(y)$$
$$F(a) \cap F(y) \neq \emptyset \neq F(b) \cap F(x),$$

and these are jointly incompatible for intervals of a linearly ordered set. Therefore (2.1) holds and $(X, \prec)$ is an interval order. □

Let $K_{1,3}$ be the bipartite graph (reflexive in the present discussion) pictured as graph II in Fig. 3.1. The Hasse diagram of its interval representation is shown as diagram II in Fig. 2.1. It follows from the preceding chapter that $(X, \prec)$ is a semiorder if and only if it is an interval order that includes no restriction that is isomorphic to diagram II's poset. The corresponding result for indifference graphs is

Theorem 3 (Roberts). *A graph is an indifference graph if and only if it is an interval graph that has no subgraph isomorphic to* $K_{1,3}$.

Basic Ties to Interval Orders

Proof. If $(X, \sim)$ is an indifference graph then, as already noted, it has no subgraph isomorphic to $K_{1,3}$. Conversely, suppose $(X, \sim)$ is an interval graph no subgraph of which is isomorphic to $K_{1,3}$. Let F be a representation of $(X, \sim)$, define $\prec$ by $x \prec y$ if $F(x) <_0 F(y)$, and suppose $(X, \prec)$ is not a semiorder. Then (2.2) of Section 2.1 must fail, say with

$$a \prec b \prec c, \quad x \sim a, \quad x \sim c,$$

and clearly $x \sim b$ also. But then the subgraph of $(X, \sim)$ on $\{a, b, c, x\}$ is a copy of $K_{1,3}$. Hence $(X, \prec)$ must be a semiorder so, by Theorem 2, $(X, \sim)$ is an indifference graph. □

Although there is only one interval graph that naturally corresponds to a given interval order, namely the symmetric complement of the interval order, a number of different interval orders can have the same symmetric complement. This is what makes the relationship between interval orders and interval graphs interesting. We shall study this relationship in depth later in this chapter and further on into the book, but will note a few more basic structural results before we look at alternative characterizations of interval graphs and indifference graphs in Sections 3.4 and 3.5.

We shall say that an interval order $(X, \prec)$ *agrees with* an interval graph $(X, \sim)$ if $\sim = sc(\prec)$. Given that $(X, \prec)$ agrees with $(X, \sim)$, it is immediate from the definition that

A is an antichain in $(X, \prec) \Leftrightarrow$
A is a clique in $(X, \sim)$;
A is a chain in $(X, \prec) \Leftrightarrow$
A is an independent set in $(X, \sim)$.

Consequently, every interval order that agrees with an interval graph $(X, \sim)$ has precisely the same set of maximal antichains, that is, the maximal cliques in $(X, \sim)$; but, because different orderings for different agreeing interval orders can rearrange elements in chains, the maximal chain sets (unordered) but not the orderings within these sets remain invariant over the family of agreeing interval orders.

It follows from Corollary 2.1 that every interval order that agrees with a finite interval graph has the same magnitude. We therefore define this value as the *magnitude* of the interval graph. To summarize:

Corollary 1. *Let $\mathscr{C}$ be the set of maximal cliques of an interval graph. Then $\mathscr{C}$ is the set of maximal antichains of every interval order that agrees with the interval graph, and if the interval graph is finite then $|\mathscr{C}|$ is its magnitude.*

3.3 LINEARLY ORDERED MAXIMAL CLIQUES

This section considers linear orderings of the set $\mathscr{C}$ of maximal cliques of $(X, \sim)$ that correspond to its agreeing interval orders. It turns out that each ordering of $\mathscr{C}$ satisfies the property that, for every $x \in X$, the maximal cliques that contain x form an interval in $\mathscr{C}$.

For finite interval graphs, this property is tantamount to the consecutive-1's-property of Fulkerson and Gross (1965). We illustrate both for the magnitude-7 interval graph that corresponds to the interval order in Fig. 2.5. The top part of Fig. 3.2 shows the natural order $<_c$ on the maximal antichains of the interval order in Fig. 2.5 along with the intervals in $(\mathscr{C}, <_c)$ that contain each point in X. The lower part of the figure translates the upper part into a matrix whose rows are ordered by $<_c$. There is a 1 in cell (p, q) if $q \in p$, and 0 otherwise. Since the 1's in each column are adjacent to each other with no intervening 0's, the matrix has the consecutive-1's-property. Matrices with this

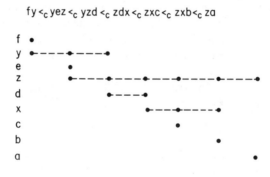

ORDER OF MAXIMAL ANTICHAINS, AND INTERVALS

	f	y	e	z	d	x	c	b	a
fy	1	1	0	0	0	0	0	0	0
yez	0	1	1	1	0	0	0	0	0
yzd	0	1	0	1	1	0	0	0	0
zdx	0	0	0	1	1	1	0	0	0
zxc	0	0	0	1	0	1	1	0	0
zxb	0	0	0	1	0	1	0	1	0
za	0	0	0	1	0	0	0	0	1

CONSECUTIVE – 1's-MATRIX FOR ROW ORDERING
OF MAXIMAL CLIQUES

Figure 3.2

property are sometimes called Petrie matrices (Shuchat, 1984), after the archaeologist Flinders Petrie.

The construction of $(\mathscr{C}, <_c)$ from a particular interval order $(X, \prec)$ that agrees with a finite interval graph $(X, \sim)$ is quite simple. One description of the process is described in the last few paragraphs of Section 2.3. Here is another. Given $(X, \prec)$, let $Y = (X^-/\sim^-) \cup (X^+/\sim^+)$ and let $\prec_0$ be the conjoint linear order on Y (Theorems 1.2, 2.3), where the notation is as in Section 2.2. For each $a \in X^-/\sim^-$ let

$$C(a) = \{x \in X : [x^-] \leqq_0 a \leqq_0 [x^+]\},$$

where square brackets denote equivalence classes in $X^-/\sim^-$ and $X^+/\sim^+$ that contain the enclosed elements. Then

$$\mathscr{C} = \{C(a) : a \in X^-/\sim^-\}$$

and

$$C(a) <_c C(b) \Leftrightarrow a \prec_0 b.$$

It is easily verified that $(\mathscr{C}, <_c)$ is a linearly ordered set on the maximal cliques in $(X, \sim)$ and that, for every x, $\{C \in \mathscr{C} : x \in C\}$ is an interval in $(\mathscr{C}, <_c)$.

We shall refer to $(\mathscr{C}, <_c)$ as the *clique ordering* for $(X, \sim)$ induced by the agreeing interval order $(X, \prec)$. Its definition for finite X is given immediately above. Its definition for X of arbitrary cardinality is more involved because of things that can happen with the sets

$$C(a) = \{x \in X : [x^-] \leqq_0 a \leqq_0 [x^+]\}$$

when X is infinite. First, it is necessary to consider $a \in X^+/\sim^+$ as well as $a \in X^-/\sim^-$ since otherwise some maximal cliques might be missed. However, some $a \in X^+/\sim^+$ may not be needed since their $C(a)$ duplicate maximal cliques identified by elements in $X^-/\sim^-$. We shall refer to such a as redundant points.

Second, it can happen that $C(a)$ is not a *maximal* clique. In this case we refer to a as deficient.

Third, there can be maximal cliques in $(X, \sim)$ that are equal to no $C(a)$. This occurs if and only if $(Y, \prec_0)$ has what we shall call complex open gaps.

The process described below defines these terms precisely and then constructs $(\mathscr{C}, <_c)$ by adding points to $(Y, \prec_0)$ in the complex open gaps and partly ignoring redundant and deficient points. Our general construction of $(\mathscr{C}, <_c)$ follows.

Let $(X, \prec)$ be an interval order that agrees with the interval graph $(X, \sim)$. Using the notation of Section 2.2, let $\prec_0$ be the conjoint weak order of $(X, \prec)$ on $X^+ \cup X^-$, and let $Y = (X^-/\sim^-) \cup (X^+/\sim^+)$ so that $(Y, \prec_0)$ is a linearly ordered set (Theorems 1.2 and 2.3).

An element $b \in X^+/\sim^+$ is *redundant* if there is an $a \in X^-/\sim^-$ such that $a \prec_0 b$ and nothing in Y lies between a and b. An element $a \in X^-/\sim^-$ is *deficient* if:
  (i) $\{c \in Y: a \prec_0 c\}$ has no minimal element; and
  (ii) there is a $b \in Y$ such that $a \prec_0 b$ and, for all $x \in X, [x^-] \leqq_0 a \prec_0 [x^+] \Rightarrow b \prec_0 [x^+]$.
Similarly, $b \in X^+/\sim^+$ is *deficient* if:
  (iii) $\{c \in Y: c \prec_0 b\}$ has no maximal element; and
  (iv) there is an $a \in Y$ such that $a \prec_0 b$ and, for all $x \in X, [x^-] \prec_0 b \leqq_0 [x^+] \Rightarrow [x^-] \prec_0 a$.
Finally, $(Y, \prec_0)$ has a *complex open gap* (Y_1, Y_2) if:
  (v) $\{Y_1, Y_2\}$ is a partition of Y and $Y_1 \prec_0 Y_2$;
  (vi) for every $a \in Y_1$ there is an $x \in X$ such that $a \prec_0 [x^-] \in Y_1$ and $[x^+] \in Y_2$; and
  (vii) for every $b \in Y_2$ there is an $x \in X$ such that $[x^-] \in Y_1$, $[x^+] \in Y_2$ and $[x^+] \prec_0 b$.

These definitions are illustrated in Fig. 3.3. Note that the definition of deficiency implies for (i) and (ii) that there is an infinite decreasing chain between a and b (see last part of Theorem 2.3), and that if (Y_1, Y_2) is a complex open gap then Y_1 has no maximal element and Y_2 has no minimal element.

Given $(Y, \prec_0)$, let $\mathcal{O}$ denote its set of complex open gaps, and extend $\prec_0$ linearly to $\prec_0^*$ on $Y^* = Y \cup \mathcal{O}$ as follows:

$$a \prec_0^* b \text{ if either } a, b \in Y \text{ and } a \prec_0 b;$$

$$\text{or } a \in Y, b = (Y_1, Y_2) \in \mathcal{O}, \text{ and } a \in Y_1;$$

$$\text{or } a = (Y_1, Y_2) \in \mathcal{O}, b \in Y, \text{ and } b \in Y_2;$$

$$\text{or } a = (Y_1, Y_2) \in \mathcal{O}, b = (Z_1, Z_2) \in \mathcal{O},$$

$$\text{and } Y_1 \subset Z_1.$$

This simply adds a "point" between Y_1 and Y_2 for each $(Y_1, Y_2) \in \mathcal{O}$. Next, let

$$C(a) = \{x \in X: [x^-] \leqq_0^* a \leqq_0^* [x^+]\}$$

for all $a \in Y^*$, and let

$$Y^{**} = \{a \in Y^*: a \text{ is neither a redundant point nor a deficient point in } Y\}.$$

Theorem 4. *Suppose $(X, \prec)$ is an interval order that agrees with interval graph $(X, \sim)$. Let $(Y^*, \prec_0^*)$ and $\{C(a): a \in Y^{**}\}$ be defined as above. Also*

Linearly Ordered Maximal Cliques

define $<_c$ *on* $\{C(a): a \in Y^{**}\}$ *by*

$$C(a) <_c C(b) \text{ if } a \prec_0^* b.$$

Then $(\{C(a): a \in Y^{**}\}, <_c)$ *is the set* $\mathscr{C}$ *of maximal cliques in* $(X, \sim)$ *linearly ordered by* $<_c$ *and, for every* $x \in X$, *the set* $\{C \in \mathscr{C}: x \in C\}$ *of maximal cliques in* $(X, \sim)$ *that contain* x *is an interval in* $(\mathscr{C}, <_c)$.

Proof. Given the hypotheses of the theorem, it suffices to show that $\{C(a): a \in Y^{**}\}$ is precisely the set $\mathscr{C}$ of maximal cliques in $(X, \sim)$ with $C(a) \neq C(b)$ if $a, b \in Y^{**}$ and $a \neq b$. Then clearly the restriction of $\prec_0^*$ on Y^{**} is a linear order, so $<_c$ linearly orders $\mathscr{C}$, and

$$\{C \in \mathscr{C}: x \in C\} = \{C(a): a \in Y^{**} \text{ and } x \in C(a)\}$$

$$= \{C(a): a \in Y^{**} \text{ and } [x^-] \leq_0^* a \leq_0^* [x^+]\}$$

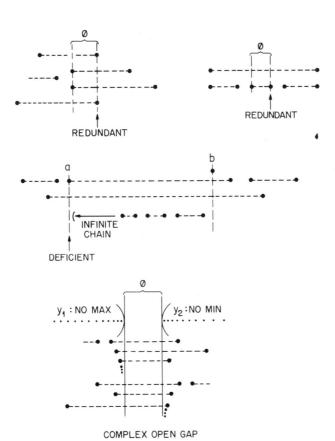

Figure 3.3

is an interval in $(\mathscr{C}, <_c)$. (Since $[x^-]$ or $[x^+]$ need not be in Y^{**}, these intervals need not be closed.)

We note first that if $b \in Y$ is deficient then $C(b)$ is not maximal, hence is not in $\mathscr{C}$. Suppose for definiteness that b is deficient and is in $X^+/\sim^+$, so conditions (iii) and (iv) in the preceding definitions apply. Since b is a right endpoint, (iii) and (iv) imply that there are $y \in X \setminus C(b)$ such that $y \sim C(b)$, and therefore clique $C(b)$ is not maximal.

We note next that if $b \in Y$ and if $C(b)$ is not maximal then b is deficient. Suppose $C(b)$ is not maximal, and for definiteness take $b \in X^+/\sim^+$. Since $C(b)$ is not maximal, there is a $y \in X \setminus C(b)$ such that $y \sim C(b)$ and $[y^+] \prec_0 b$. Let $a = [y^+]$. Then, since $y \sim C(b)$, the definition of $\prec_0$ requires $[x^-] \prec_0 a$ for every $x \in C(b)$, that is, for every x for which $[x^-] \prec_0 b \leqq [x^+]$, and this verifies (iv). If $\{c \in Y: c \prec_0 b\}$ has a maximal element, it must be a $[z^-] \in X^-/\sim^-$ (see Theorem 2.3), but then $z \in C(b)$, and $y \prec z$ since $[y^+] \prec_0 [z^-]$, contrary to $y \sim C(b)$. Hence $\{c \in Y: c \prec_0 b\}$ has no maximal element, so (iii) holds and b is deficient.

Thus far we have shown that, for all $b \in Y$,

$$C(b) \text{ is in } \mathscr{C} \Leftrightarrow b \text{ is not deficient.}$$

We show next that if $b \in Y$ and $C(b) \in \mathscr{C}$ then there is a nonredundant $a \in Y$ for which $C(a) = C(b)$. Suppose $b \in Y$, $C(b) \in \mathscr{C}$, and b is redundant. Then $b \in X^+/\sim^+$ and there is a maximal $a \prec_0 b$ with $a \in X^-/\sim^-$. It is easily seen that $C(a) = C(b)$ and, since a is not in $X^+/\sim^+$, it is not redundant.

Suppose now that C is a maximal clique in $\mathscr{C}$ that is equal to no $C(a)$ for $a \in Y$. We shall prove that $C = C(g)$ for some $g \in \mathcal{O}$. Given $C \neq C(a)$ for all $a \in Y$, for every $a \in Y$ it must be true that there is an $x \in C$ such that

$$\text{either } [x^+] \prec_0 a \text{ or } a \prec_0 [x^-],$$

for otherwise $C = C(a)$ for some a. Let

$$Y_1 = \{a \in Y: a \prec_0 [x^-] \text{ for some } x \in C\},$$
$$Y_2 = \{a \in Y: [x^+] \prec_0 a \text{ for some } x \in C\},$$

so that $Y_1 \cup Y_2 = Y$. If $Y_1 = \emptyset$, then $x \in C$ implies that there is a $y \in C$ such that $[y^+] \prec_0 [x^-]$, hence $y \prec x$, which contradicts $x \sim y$. Therefore $Y_1 \neq \emptyset$. Similarly, $Y_2 \neq \emptyset$. Moreover, it cannot be true for $x, y \in C$ that $[x^+] \prec_0 a \prec_0 b \prec_0 [y^-]$ since otherwise $x \prec y$. Therefore $Y_1 \prec_0 Y_2$.

Thus, given $C \neq C(a)$ for all $a \in Y$, there is a partition $\{Y_1, Y_2\}$ of Y for which $Y_1 \prec_0 Y_2$ such that

$$a \in Y_1 \Rightarrow a \prec_0 [x^-] \text{ for some } x \in C,$$
$$b \in Y_2 \Rightarrow [x^+] \prec_0 b \text{ for some } x \in C.$$

It is easily seen that $\mathscr{C} = C(a)$ for some $a \in Y$ if either $(Y_1, \prec_0)$ has a maximal element or $(Y_2, \prec_0)$ has a minimal element. Consequently, $(Y_1, \prec_0)$ is open above and $(Y_2, \prec_0)$ is open below: see Fig. 3.3. Suppose $x \in C$. Then the preceding implications force $[x^-] \in Y_1$ and $[x^+] \in Y^2$. For example, if $[x^-] \in Y_2$ then $[y^+] \prec_0 [x^-]$ for some $y \in C$ by the second implication, so that $y \not\sim x$ for $x, y \in C$.

It follows from definitions (v)–(vii) that (Y_1, Y_2) is a complex open gap in $(Y, \prec_0)$. Let $g = (Y_1, Y_2)$. Then by the extension of $\prec_0$ to $\prec_0^*$ on $Y^* = Y \cup \mathcal{O}$, we have $[x^-] \prec_0^* g \prec_0^* [x^+]$ for every $x \in C$. In fact, $C = C(g)$, for if $[x^-] \prec_0^* g \prec_0^* [x^+]$ for some $x \notin C$, then $x \sim C$ (since, for every $y \in C$, $[x^-] \prec_0 [y^+]$ and $[y^-] \prec_0 [x^+]$), contrary to the maximality of C.

Since two distinct maximal cliques in $\mathscr{C}$ that are not in $\{C(a): a \in Y\}$ must have different complex open gaps, for any such C there is a unique $g \in Y^{**}$ such that $C = C(g)$.

We are almost done. Thus far we know that each $C \in \mathscr{C}$ that is in $\{C(a): a \in Y\}$ has one or more nonredundant and nondeficient $a \in Y$ for which $C = C(a)$, and that each $C \in \mathscr{C}$ that is not in $\{C(a): a \in Y\}$ has exactly one $g \in \mathcal{O}$ for which $C = C(g)$.

It remains only to prove that, when $C \in \mathscr{C}$, there do not exist distinct nonredundant $a, b \in Y$ for which $C = C(a) = C(b)$. Assume $a, b \in Y$, $a \prec_0 b$, $C(a) = C(b)$, and neither a nor b is redundant. We prove that a or b is deficient, hence that C is not maximal. Since b is not redundant, $a \prec_0 d \prec_0 b$ for some $d \in Y$. Then $C(a) = C(b)$ forces $a \prec_0 [x^-] \prec_0 [x^+] \prec_0 b$ when $d \in \{[x^-], [x^+]\}$. In addition, $C(a) = C(b)$ implies either $\{a \in X^-/\sim^-$ and $b \leqq_0 [y^+]$ for every y such that $[y^-] \leqq_0 a \prec_0 [y^+]\}$ or $\{b \in X^+/\sim^+$ and $[y^-] \leqq_0 a$ for every y such that $[y^-] \prec_0 b \leqq_0 [y^+]\}$. Suppose for definiteness that the first alternative obtains:

$$a \in X^-/\sim^- \quad \text{and} \quad [y^-] \leqq_0 a \prec_0 [y^+] \Rightarrow b \leqq_0 [y^+].$$

We show that a is deficient. Since $a \prec_0 [x^+] \prec_0 [b] \leqq_0 [y^+]$ whenever $[y^-] \leqq_0 a \prec_0 [y^+]$, we need only prove that $\{c \in Y: a \prec_0 c\}$ has no minimal element. Suppose it does. Since $a \in X^-/\sim^-$, this minimal element must be a $[z^+] \in X^+/\sim^+$. However, we then have $[z^-] \leqq_0 a$, hence $z \in C(a)$, and $[z^+] \prec_0 b$, hence $z \notin C(b)$, contrary to $C(a) = C(b)$. Hence $\{c \in Y: a \prec_0 c\}$ has no minimal element and a is deficient. □

Theorem 4 shows constructively that every interval order $(X, \prec)$ that agrees with an interval graph $(X, \sim)$ induces a linear clique ordering $<_c$ on the set $\mathscr{C}$ of maximal cliques in $(X, \sim)$ such that $\{C \in \mathscr{C}: x \in C\}$ is an interval in $(\mathscr{C}, <_c)$ for every $x \in X$.

The following corollary is similar to results established by Gilmore and Hoffman (1964) and Fulkerson and Gross (1965). We say that a linear ordering of the maximal cliques of a graph is *consecutive* if, for every x in the graph, the maximal cliques that contain x form an interval in the ordering.

Corollary 2. *A reflexive graph is an interval graph if and only if there is a consecutive linear ordering of its maximal cliques.*

Proof. Suppose $(X, \sim)$ is an interval graph. Let F be a representation of $(X, \sim)$, and let $(X, \prec)$ be the interval order defined from F in the natural way. Then $(X, \prec)$ agrees with $(X, \sim)$, and the conclusion of Corollary 2 follows from Theorem 4. The converse proof follows directly from the definition of $F(x)$ as $\{C \in \mathscr{C} : x \in C\}$. □

3.4 CHARACTERIZATIONS OF INTERVAL GRAPHS

Corollary 2 presents one of the standard characterizations of interval graphs. We establish two others in this section. The first is due to Gilmore and Hoffman (1962, 1964), the second to Lekkerkerker and Boland (1962). Both are forbidden-subgraph characterizations in the sense that their conditions forbid $(X, \sim)$ to contain certain configurations.

Theorem 5 (Gilmore and Hoffman). *A reflexive graph $(X, \sim)$ is an interval graph if and only if every simple $\sim$ cycle of length four has a chord and $c(X, \sim) = (X, \not\prec)$ is a comparability graph.*

Recall from Theorem 1.7 that $(X, \not\prec)$ is a comparability graph if and only if every $\not\prec$ cycle of odd length has a triangular chord. The other condition of Theorem 5 says that $(X, \sim)$ has no subgraph isomorphic to $K_{2,2}$.

Proof. The conditions are clearly necessary for $(X, \sim)$ to be an interval graph. To prove sufficiency, assume the conditions and let $\prec$ on X with $\not\prec = \prec \cup d(\prec)$ be a partial order as guaranteed by the comparability-graph condition. Since $\sim = sc(\prec)$, Theorem 1 tells us that $(X, \sim)$ is an interval graph if $(X, \prec)$ is an interval order. If $(X, \prec)$ is not an interval order, then (2.1) must fail (Section 2.1), hence for some $a, b, x, y \in X$,

$$a \prec x, \quad b \prec y, \quad a \sim y, \quad \text{and} \quad b \sim x,$$

which entails $a \sim b$, $x \sim y$, and $|\{a, b, x, y\}| = 4$. But then $a \sim b \sim x \sim y \sim a$ is a simple four-cycle with no chord, contrary to assumption. Hence $(X, \sim)$ is an interval graph. □

Unlike the Gilmore-Hoffman characterization, the one developed by Lekkerkerker and Boland uses conditions stated directly in terms of $\sim$. We refer to $(X, \sim)$ as *triangulated* if every simple $\sim$ cycle of length four or more has a chord.

Theorem 6 (Lekkerkerker and Boland). *A reflexive graph is an interval graph if and only if it is triangulated and has no asteroidal triple.*

Prior to proving this, we note an important theorem for triangulated graphs due to Dirac (1961). A point $x \in X$ is a *simplicial point* of $(X, \sim)$ if the subgraph on $\{y \in X: y \sim x\} \cup \{x\}$ is complete. In addition, a subset $Y \subseteq X$ is a *separator* for nonadjacent $(x \not\sim y)$ points x and y in graph $(X, \sim)$ if x and y are in different components of $(X \setminus Y, \sim)$.

Theorem 7 (Dirac). *Every finite triangulated graph has a simplicial point. Moreover, every finite triangulated graph that is not complete has two nonadjacent simplicial points.*

Remark 1. Since the indifference graph $(\mathbb{Z}, \sim)$ for which $n \sim m \Leftrightarrow |n - m| \leq 1$ has no simplicial point, the theorem can fail for infinite triangulated graphs.

Remark 2. Let $\mathscr{C}$ be the set of maximal cliques of $(X, \sim)$. Theorem 7 says that if X is finite and $(X, \sim)$ is triangulated, then some $x \in X$ is in exactly one $C \in \mathscr{C}$, and if $|\mathscr{C}| > 1$ then there are two such points in different members of $\mathscr{C}$.

The following proof of Theorem 7 is patterned after Golumbic (1980, p. 83).

Proof. Let $(X, \sim)$ be a finite triangulated graph. We show first that if Y is a minimal separator for nonadjacent points x_1 and x_2, then either $Y = \emptyset$ or $(Y, \sim)$ is complete. Assume Y is minimal and has two or more points. Let $(X_1, \sim)$ and $(X_2, \sim)$ be the components of $(X \setminus Y, \sim)$ that respectively contain x_1 and x_2. Then, since Y is minimal, each $y \in Y$ has $y \sim a$ for some $a \in X_1$ and $y \sim b$ for some $b \in X_2$. Therefore, for any distinct y and z in Y there are minimum-length paths $(y, a_1, \ldots, a_r, z)$ with all $a_i \in X_1$ and $(z, b_1, \ldots, b_t, y)$ with all $b_i \in X_2$. It follows that $(y, a_1, \ldots, a_r, z, b_1, \ldots, b_t, y)$ is a simple cycle of length four or more. Since $(X, \sim)$ is triangulated, this cycle has a chord which, by choice of paths, must be either $y \sim z$ or an $a_i \sim b_j$. The latter is false since $(X_1, \sim)$ and $(X_2, \sim)$ are different components of $(X \setminus Y, \sim)$, so $y \sim z$. Hence $(Y, \sim)$ is complete.

The proof now proceeds by induction on $|X|$. If $(X, \sim)$ is complete, the result is obvious, so assume that $(X, \sim)$ has nonadjacent x_1 and x_2 and that the theorem is true for graphs with fewer points. With Y, X_1 and X_2 as above, let $Z = (X_1 \cup Y, \sim)$. By induction, either Z has two nonadjacent simplicial points, at least one of which is in X_1 since either Y is empty or $(Y, \sim)$ is complete, or Z is complete, in which case every point in X_1 is simplicial in Z. Moreover, $\{t \in X: t \sim a \text{ for some } a \in X_1\} \subseteq X_1 \cup Y$. Therefore, some simplicial point in Z that lies in X_1 is simplicial in $(X, \sim)$. Similarly, X_2 contains a simplicial point of $(X, \sim)$. □

We now turn to a proof of Theorem 6, patterned after Lekkerkerker and Boland (1962) but using Corollary 2 for maximal cliques where they use real representations. We say that a simplicial point x of $(X, \sim)$ is *strong* if $(X \setminus (\{y: y \sim x\} \cup \{x\}), \sim)$ is connected, and is *weak* otherwise. In addition, a maximal clique in an ordering of maximal cliques is an *end clique* if it is first or last in the ordering.

Proof of Theorem 6. Since the conditions of the theorem are easily seen to be necessary for $(X, \sim)$ to be an interval graph, we turn to the sufficiency proof. It will be assumed that $(X, \sim)$ is a triangulated reflexive graph that is *not* an interval graph. It will then be shown that $(X, \sim)$ has an asteroidal triple.

Let $(X, \sim)$ be a triangulated reflexive graph that is not an interval graph. By Theorem 5, it has an odd-length $\nsim$ cycle with no triangular chord. Therefore some *finite* subgraph of $(X, \sim)$ is not an interval graph.

Assume henceforth that Y is a minimum-cardinality subset of X such that $(Y, \sim)$ is not an interval graph. Clearly, Y is finite with more than three points, $(Y, \sim)$ is reflexive, triangulated, connected, and not complete, every nonempty proper subgraph of $(Y, \sim)$ is an interval graph, and no two distinct simplicial points of $(Y, \sim)$ are adjacent. Let $\mathscr{C}$ and $\mathscr{C}(x)$ for $x \in Y$ denote respectively the sets of maximal cliques in $(Y, \sim)$ and $(Y \setminus \{x\}, \sim)$.

By Theorem 7, $(Y, \sim)$ has at least two simplicial points. If it has three *strong* simplicial points, then it has an asteroidal triple since the deletion of all points adjacent to any one of these leaves a connected subgraph containing the other two. We prove that other possibilities yield contradictions.

Suppose first that $(Y, \sim)$ has exactly two simplicial points. Let x be one of these with maximal clique $\{x\} \cup A$ ($x \notin A$) in Y. By Corollary 2, let $<_c$ be a consecutive linear ordering of $\mathscr{C}(x)$. One or more C in $\mathscr{C}(x)$ have $A \subseteq C$. If none of these is an end clique in $(\mathscr{C}(x), <_c)$, then $(Y, \sim)$ has at least two simplicial points besides x (points in the end cliques that are not in others), a contradiction. Hence some $C \supseteq A$ in $\mathscr{C}(x)$ is an end clique. But then $A \cup \{x\}$ can be added at that end—or, if A itself is in $\mathscr{C}(x)$, just add x to A—to obtain a consecutive linear ordering of $\mathscr{C}$. But then $(Y, \sim)$ is an interval graph by Corollary 2, a contradiction.

Assume henceforth that $(Y, \sim)$ has at least three simplicial points but not three strong simplicial points. Throughout the rest of the proof let x be a *weak* simplicial point of $(Y, \sim)$ with maximal clique $\{x\} \cup A, x \notin A$, and let $X_1, \ldots, X_m$ ($m \geq 2$) be the components of $(Y \setminus (\{x\} \cup A), \sim)$. Let $<_c$ be a consecutive linear ordering of $\mathscr{C}(x)$. If either $A \in \mathscr{C}(x)$ or $X_i \sim A$ for some i, then we get a consecutive linear ordering of $\mathscr{C}$ from $<_c$ [add x to A in the first case; insert $\{x\} \cup A$ at one end of the cliques in $\mathscr{C}(x)$ that contain points in X_i in the second case], contrary to hypothesis.

For definiteness assume henceforth that X_1 contains one or more $y \sim A$ and one or more z with $z \nsim a$ for some $a \in A$. Such an X_i occurs when $A \notin \mathscr{C}(x)$ and, for all i, it is false that $X_i \sim A$. Let

$$Y_1 = \{x\} \cup A \cup X_1; \qquad \mathscr{C}_1 = \text{set of maximal cliques in } (Y_1, \sim).$$

Also let $<_1$ be a consecutive linear ordering of $\mathscr{C}_1$, as guaranteed by Corollary

Characterizations of Interval Graphs

2. Since *x is a strong simplicial point of* $(Y_1, \sim)$, it is easily seen that $\{x\} \cup A$ is an end clique in $(\mathscr{C}_1, <_1)$. Without loss of generality, suppose $\{x\} \cup A$ is last under $<_1$, and let z be a point in X_1 that is in only the first maximal clique of $(\mathscr{C}_1, <_1)$, with $z \not\sim a$ for some $a \in A$. The existence of z is assured by the choice of X_1 and the maximality of cliques in $\mathscr{C}_1$. Now, given $(\mathscr{C}(x), <_c)$ as in the preceding paragraph, assume that the clique in $\mathscr{C}_1$ that contains z (which is also a clique in $\mathscr{C}(x)$) precedes by $<_c$ the clique or cliques in $\mathscr{C}(x)$ that properly include A. If this is not true for $<_c$, simply replace it by its dual. Because A is not in $\mathscr{C}(x)$, and X_1 is a component of $(Y \setminus (\{x\} \cup A), \sim)$, the cliques in $\mathscr{C}(x)$ that contain the points in X_1 must be an interval in $(\mathscr{C}(x), <_c)$. Let $\mathscr{D}$ be this interval. All $D \in \mathscr{D}$ have points in X_1 but no points in $X_2, \ldots, X_m$, which lie in the cliques in $\mathscr{C}(x)$ that precede and follow $\mathscr{D}$. Moreover, with C_z the unique maximal clique in $\mathscr{C}(x)$ and $\mathscr{C}_1$ that contains z, the only points in A that can be in cliques of $\mathscr{C}(x)$ that precede $\mathscr{D}$ are points in $A \cap C_z$. This follows from the choice of z.

Finally, form $(\mathscr{C}, <'_c)$ by replacing $\mathscr{D}$ within $(\mathscr{C}(x), <_c)$ by $(\mathscr{C}_1, <_1)$, noting that $\mathscr{C}_1$ consists precisely of the cliques in $\mathscr{D}$ plus $\{x\} \cup A$. The fact that $\{x\} \cup A$ is last under $<_1$ plus the later observations in the preceding paragraph imply that $<'_c$ is a consecutive linear ordering of $\mathscr{C}$. But then $(Y, \sim)$ is an interval graph according to Corollary 2, and this contradiction completes the proof. $\square$

Lekkerkerker and Boland (1962) have determined the minimal graphs that are not interval graphs: see Fig. 3.4. The graphs in the upper left are the simple cycles ($n \geq 4$) with no chords. The others are triangulated and have asteroidal triples. Each of these has exactly three simplicial points, all of which are strong. Their Theorem 4 says that a reflexive graph is an interval graph if and only if it has no subgraph that is isomorphic to one of the graphs shown in the figure.

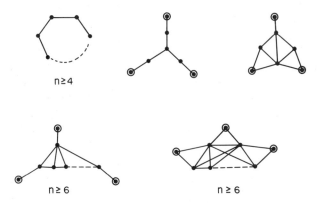

Figure 3.4 Minimal noninterval graphs.

3.5 CHARACTERIZATIONS OF INDIFFERENCE GRAPHS

We say that a semiorder $(X, \prec)$ *agrees with* an indifference graph $(X, \sim)$ if $\sim \, = sc(\prec)$. By Theorem 2, $(X, \sim)$ is an indifference graph if and only if it has an agreeing semiorder. By Theorem 3, $(X, \sim)$ is an indifference graph if and only if it is an interval graph (has an agreeing interval order) that has no subgraph isomorphic to $K_{1,3}$.

Other characterizations of indifference graphs are similar to the results for interval graphs in the two preceding sections. We consider clique orderings first.

Recall that a linear ordering $<_c$ of the set $\mathscr{C}$ of maximal cliques in $(X, \sim)$ is *consecutive* if, for every x in X, $\{C \in \mathscr{C}: x \in C\}$ is an interval in $(\mathscr{C}, <_c)$. Such a $<_c$ will be said to be *proper* if, for all x and y in X, there do not exist C' and C'' in $\mathscr{C}$ that contain x such that $C' <_c \{C \in \mathscr{C}: y \in C\} <_c C''$.

Theorem 8. *Suppose $(X, \prec)$ is a semiorder that agrees with indifference graph $(X, \sim)$. Let $(\mathscr{C}, <_c)$ be defined in Theorem 4. Then $<_c$ is a proper consecutive linear ordering of the set $\mathscr{C}$ of maximal cliques in $(X, \sim)$.*

Proof. In view of Theorem 4 it is only necessary to check that $<_c$ is proper. To the contrary, suppose that there are $x, y \in X$ and $a', a'' \in Y^{**}$ such that, with $C(a) = \{u \in X: [u^-] \leq_0^* a \leq_0^* [u^+]\}$ for all $a \in Y^*$,

$$x \in C(a') \cap C(a'')$$

and

$$[y \in C(b), b \in Y^{**}] \Rightarrow a' \prec_0^* b \prec_0^* a''.$$

We show that there are $z, w \in X$ such that $x \sim z \prec y$ and $y \prec w \sim x$, which contradict the hypothesis that $(X, \prec)$ is a semiorder: see (2.2).

Consider $a' \in Y^{**}$, for which $a' \prec_0^* [y^-]$ since otherwise $y \in C(a')$. If $a' = [z^+]$ for some $z \in X$, then $x \sim z \prec y$. If $a' = [t^-]$ for some $t \in X$ then, since a' is not deficient (it is in Y^{**}), either $\{c \in Y: a' \prec_0 c\}$ has a minimal element in Y, or $\{c \in Y: a' \prec_0 c\}$ has no minimal element and for every $d \in Y$ such that $a' \prec_0 d$ there is a $z \in X$ such that $[z^-] \leq_0 a' \prec_0 [z^+] \leq_0 d$. In the first case the minimal element in $\{c \in Y: a' \prec_0 c\}$ must be a $[p^+]$ by Theorem 2.3, so we have $[p^-] \leq_0 a' \prec_0 [p^+] \prec_0 [y^-]$, hence $x \sim p \prec y$. In the second case, with no minimal in $\{c \in Y: a' \prec_0 c\}$, take $a' \prec_0 c \prec_0 [y^-]$, then get $z \in X$ such that $[z^-] \leq_0 a' \prec_0 [z^+] \leq_0 c \prec_0 [y^-]$, hence $x \sim z \prec y$. Finally, if a' is a complex open gap (Y_1, Y_2), then $[x^-] \prec_0^* a' \prec_0^* [y^-]$ and there is a z in X such that $[z^-] \in Y_1, [z^+] \in Y_2$, and $[z^+] \prec_0 [y^-]$, so again $x \sim z \prec y$. Hence $x \sim z \prec y$ for some $z \in X$.

The proof for $a'' \in Y^{**}$ and $[y^+] \prec_0^* a''$ is similar and gives $y \prec w \sim x$ for some w in X. □

The proof of the following corollary is left to the reader.

Corollary 3. *A reflexive graph is an indifference graph if and only if there is a proper consecutive linear ordering of its maximal cliques.*

The obvious parallels for indifference graphs to the Gilmore-Hoffman and Lekkerkerker-Boland characterizations of interval graphs follow immediately from Theorems 3, 5, and 6. A reflexive graph is an indifference graph ⇔ its complement is a comparability graph and it has no subgraph isomorphic to $K_{2,2}$ or $K_{1,3}$ ⇔ it is triangulated, has no asteroidal triple, and has no subgraph isomorphic to $K_{1,3}$.

Roberts (1969) sharpens the latter result substantially by reducing the prohibition against asteroidal triples to graphs G_1 and G_2 shown in the lower part of Fig. 3.5. These are the $n = 6$ graphs in the lower part of Fig. 3.4. For convenience, I shall say that $(X, \sim)$ has no *astral triple* if it has no subgraph isomorphic to $K_{1,3}$, G_1 or G_2.

Theorem 9 (Roberts). *A reflexive graph is an indifference graph if and only if it is triangulated and has no astral triple.*

Proof. The conditions are easily seen to be necessary. To prove sufficiency, assume that $(X, \sim)$ is not an indifference graph but is reflexive, triangulated, and has no copy of $K_{1,3}$. We shall prove that some subgraph of $(X, \sim)$ is isomorphic to G_1 or G_2.

It follows from the paragraph after Corollary 3 that there is a finite minimum-cardinality Y in X such that $(Y, \sim)$ is not an indifference graph. Given such a $(Y, \sim)$, it is reflexive, triangulated, connected, has no copy of $K_{1,3}$, and is not complete. Moreover, no two distinct simplicial points of $(Y, \sim)$ are adacent.

The proof of Theorem 6 shows that $(Y, \sim)$ has exactly three strong simplicial points, say x, y, and z, and no weak simplicial points. Let $p(x, y)$

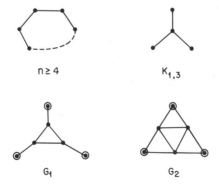

Figure 3.5 Minimal nonindifference graphs.

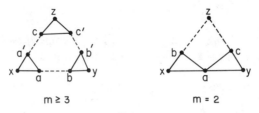

Figure 3.6

be a shortest path in $(Y, \sim)$ no point of which is adacent to z, and define $p(x, z)$ and $p(y, z)$ similarly. Also let m be the length of the shortest of these three paths. We distinguish two cases.

CASE 1. $m \geq 3$. The left picture of Fig. 3.6 illustrates this case, with $a \neq b$, $a' \neq c$, and $b' \neq c'$. We allow the possibility that $a = a'$, or $b = b'$, or $c = c'$. If one of these three equalities does not hold, then there must be an edge within the "interior" of the graph since it is triangulated. Moreover, some such edge will have exactly one of a through c' as an end point, and it is easily seen that the graph can be reduced to another graph that is not an indifference graph, contrary to the assumed minimality of Y. Hence $a = a'$, $b = b'$, and $c = c'$. Then, if any of $p(x, y)$, $p(x, z)$, and $p(y, z)$ has length exceeding 3, further interior edges caused by the triangulated property allow a similar reduction, so we conclude that each path has length 3. Hence $(Y, \sim)$ is isomorphic to G_1.

CASE 2. $m = 2$. Assume for definiteness that $p(x, y)$ has length 2 as shown on the right of Fig. 3.6. If $a \sim t$ for any $t \notin \{a, b, c, x, y\}$, then we get a copy of $K_{1,3}$ with $\{a, x, y, t\}$. Hence $a \not\sim t$ for all $t \notin \{a, b, c, x, y\}$, which, by the triangulated property, forces $b \sim c$. If there are other points between b and z or between c and z, then the graph can be reduced to a smaller nonindifference graph (z gets cut out), contrary to minimality. Hence $(Y, \sim)$ in this case is isomorphic to G_2. □

An alternative proof of Theorem 9 is readily obtained from Fig. 3.4 by showing that G_1 and G_2 are the only minimal graphs with asteroidal triples that do not have $K_{1,3}$ as a subgraph. The proof just sketched does not rely on Fig. 3.4.

3.6 UNIQUE AGREEMENT

Earlier theorems in this chapter reveal intimate connections between interval graphs and interval orders, and between indifference graphs and semiorders. The purpose of this section is to characterize graphs for which this relationship is essentially one-to-one, that is, interval graphs that have uniquely agreeing interval orders up to duality of the ordering relation. This will be done for finite interval graphs and for arbitrary indifference graphs.

A few special cases will be disposed of before we proceed. Since $sc(\prec) = scd(\prec)$, $(X, \prec)$ agrees with interval graph $(X, \sim)$ if and only if $(X, d(\prec))$ agrees with $(X, \sim)$. Consequently, every interval graph that is not complete has at least two agreeing interval orders, and an interval graph has exactly one agreeing interval order if and only if it is complete.

Since the complete-graph case is trivial, we shall focus on graphs that are not complete. If $(X, \sim)$ is an interval graph that is not connected, it is easily seen that it has an agreeing interval order unique up to duality if and only if it has exactly two components each of which is complete. For, if it has more than two components, permutations of components generate more than two agreeing interval orders, and if it has only two components at least one of which is not complete, then it has at least four agreeing interval orders.

Note also that if $(X, \sim)$ has universal points $u \sim X$ but is not complete, removal of its universal points has no effect on the number of agreeing interval orders. In particular, if $(X, \sim)$ has a universal point and the removal of all such points leaves a graph with exactly two components each of which is complete, then it has an agreeing interval order unique up to duality.

We shall focus on connected graphs with no universal points and refer to these as *nonuniversally connected* graphs. In addition, an interval graph (indifference graph) that has a uniquely agreeing interval order (semiorder) up to duality will be said to be *uniquely orderable*. Our first result, due to Roberts (1971), omits the nonuniversality condition and does not restrict the cardinality of X.

Theorem 10 (Roberts). *Every connected indifference graph is uniquely orderable.*

Proof. Suppose $(X, \sim)$ is a connected indifference graph that is *not* uniquely orderable. Then there must be $a, b, x, y \in X$ and semiorders $\prec_1$ and $\prec_2$ that agree with $(X, \sim)$ such that

$$x \prec_1 y \quad \text{and} \quad x \prec_2 y,$$
$$a \prec_1 b \quad \text{and} \quad b \prec_2 a.$$

There are paths between each two elements in $\{a, b, x, y\}$. Let Y consist of the points in these paths (one for each pair). Then Y is finite, $(Y, \sim)$ is a connected indifference graph, and the restrictions of $\prec_1$ and $\prec_2$ on Y are semiorders that agree with $(Y, \sim)$. We work with $(Y, \sim)$ henceforth and obtain a contradiction.

By (2.1) of Section 2.1, either $x \prec_1 b$ or $a \prec_1 y$. Suppose $x \prec_1 b$. Then $x \not\prec_2 b$, so either $x \prec_2 b \prec_2 a$ (hence $x \not\prec_1 a$ and either $x \prec_1 a \prec_1 b$ or $a \prec_1 x \prec_1 b$) or $b \prec_2 x \prec_2 y$ (hence $b \not\prec_1 y$ and either $x \prec_1 b \prec_1 y$ or $x \prec_1 y \prec_1 b$). A similar analysis with $a \prec_1 y$ instead of $x \prec_1 b$ shows that there are distinct u, v, and w in $\{a, b, x, y\}$ such that the middle element in $\prec_1$ on $\{u, v, w\}$ is different than the middle element in $\prec_2$ on $\{u, v, w\}$. But then a simple path analysis yields a contradiction. For example, if $u \prec_1 v \prec_1 w$ and $w \prec_2 u \prec_2 v$, then connectedness along with a representation theorem such as Theorem 2.7 or 2.9 shows from $u \prec_1 v \prec_1 w$ that there is a $\sim$ path between v and w in $(Y, \sim)$ that has no element adjacent to u. However, $w \prec_2 u \prec_2 v$ implies that every path between v and w has an element adjacent to u, so we obtain a contradiction. □

The smallest nonuniversally connected interval graph which is not uniquely orderable is the five-point graph shown on the left in Fig. 3.7. The reason why it is not uniquely orderable is that the order between x and y, as shown in the representation on the right in the figure, is independent of the order used for $\{a, b, c\}$. Hanlon (1982) regards the two orders with $a \prec b \prec c$ (one with $x \prec y$, the other with $y \prec x$) as equivalent since they are isomorphic. Our approach is different since it is "labeled" rather than "unlabeled." Hanlon characterizes finite interval graphs that are uniquely orderable up to duality *and* up to isomorphism between agreeing interval orders.

The first of our two characterizations of uniquely orderable finite interval graphs is based on Hanlon's notion of buried subgraphs. For any $\emptyset \subset A \subseteq X$ let

$$K(A) = \{x \in X : x \sim A\},$$

the set of points adjacent to every point in A. We say that $B \subseteq X$ is a *buried subgraph* of an interval graph $(X, \sim)$ if:

(i) there are $x, y \in B$ with $x \not\sim y$;
(ii) $K(B) \neq \emptyset$ and $B \cap K(B) = \emptyset$; and
(iii) every path between a point in B and a point in $X \setminus B$ contains a point in $K(B)$.

We shall refer to $K(B)$ in (ii) as the *K-set* of B. Part (iii) of the definition says that we can "escape from" B only by going through its K-set. There is one buried subgraph in Fig. 3.7, namely $\{x, y\}$. Its K-set is $\{c\}$. Every path from $\{x, y\}$ to $\{a, b\}$ goes through c.

Theorem 11 (Hanlon). *A nonuniversally connected finite interval graph is uniquely orderable if and only if it has no buried subgraph.*

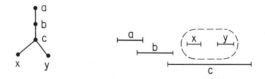

Figure 3.7 An interval graph with four agreeing orders.

This will be proved conjointly with the next theorem, which is based on a binary relation Q defined on $c(\sim) = \{xy \in X \times X: x \not\sim y\}$ as follows:

$$Q = t(Q_0),$$

$abQ_0 xy$ if $ab, xy \in c(\sim)$, $a \sim x$, and $b \sim y$.

If $abQ_0 xy$, then $a \prec b$ in an agreeing interval order if and only if $x \prec y$ in that order. If $abQxy$ for the transitive closure Q of Q_0, then $a \prec b \Leftrightarrow x \prec y$ for every interval order that agrees with $(X, \sim)$. Thus Q shows how certain pairs with $x \not\sim y$ must be ordered in any agreeing interval order, given an orientation for other pairs whose points are not adjacent.

Since Q_0 is reflexive and symmetric, Q is an equivalence relation on $c(\sim)$. Since $x \not\sim y$ for some $x, y \in X$ when $(X, \sim)$ is nonuniversally connected, and since not($xyQyx$), the set $c(\sim)/Q$ of equivalence classes has at least two members. The following theorem says that unique orderability is tantamount to a choice of orientation for one pair $\{a, b\}$ with $a \not\sim b$ determining the orientation for every such pair.

Theorem 12. *A nonuniversally connected finite interval graph $(X, \sim)$ is uniquely orderable if and only if $|c(\sim)/Q| = 2$.*

Proofs. Let $(X, \sim)$ be a nonuniversally connected finite interval graph. We shall prove that

$(|c(\sim)/Q| = 2) \Rightarrow$ unique orderability,

unique orderability $\Rightarrow$ no buried subgraph,

no buried subgraph $\Rightarrow (|c(\sim)/Q| = 2)$.

The first two implications are easily verified. If $|c(\sim)/Q| = 2$, then one equivalence class must be the dual of the other (xy in one $\Leftrightarrow yx$ in the other) and, by remarks preceding Theorem 12, an orientation for one $\{x, y\}$ with $x \not\sim y$ determines an orientation for all such pairs in an agreeing interval order. Moreover, if $(X, \sim)$ has a buried subgraph B and $(X, \prec)$ agrees with $(X, \sim)$, then another agreeing order that differs from both $\prec$ and $d(\prec)$ (because B contains $x \not\sim y$, and because the graph is not universally connected) is obtained from $\prec$ by replacing $(B, \prec)$ with $(B, d(\prec))$. In other words, reorient dually within B.

To prove the final implication, assume that $(X, \sim)$ has no buried subgraph. We wish to show that $|c(\sim)/Q| = 2$. Let $(X, \prec)$ be an interval order that agrees with $(X, \sim)$, and let (f, ρ) be a representation of $(X, \prec)$ with $\rho > 0$. Choose x and y so that x minimizes $f(a) + \rho(a)$ and y maximizes $f(a)$. Since $(X, \sim)$ is not complete, $x \prec y$ and $f(x) + \rho(x) < f(y)$. We shall prove that $abQxy$ for all a, b that have $a \prec b$, so that one class in $c(\sim)/Q$ is $\{ab: a \prec b\} = \{ab: f(a) + \rho(a) < f(b)\}$. The other class must then be $\{ab: b \prec a\}$, and the two together exhaust $c(\sim)$.

We shall suppose that not($abQxy$) for some $a \prec b$, and show that this implies that $(X, \sim)$ has a buried subgraph. It then follows that, if $(X, \sim)$ has no buried subgraph, $|c(\sim)/Q| = 2$.

Assume not($abQxy$) for some $a \prec b$. Choose such an ab so that $f(a) + \rho(a)$ is minimized and, given this a, $f(b)$ is maximized. Let

$$C = \{c \in X: f(c) + \rho(c) < f(a) + \rho(a)\},$$

$$D = \{d \in X: f(b) < f(d)\}.$$

Then, since either $x \in C$ or $y \in D$, $C \cup D \neq \emptyset$.

By the choice of ab,

$$xyQ\{cb, ad, cd\}$$

for all $c \in C$ and all $d \in D$. Since not($abQxy$), every path from a to $c \in C$ must have a point adjacent to b (else $abQcbQxy$), and every path from b to $d \in D$ must have a point adjacent to a. Let

$$p = \max\{f(c) + \rho(c): c \in C\}, \quad q = \min\{f(d): d \in D\},$$

with $p = -\infty$ if $C = \emptyset$, and $q = \infty$ if $D = \emptyset$. We then have the picture shown in Fig. 3.8 if both C and D are nonempty. Every path from a to C requires a point like r, and every path from b to D requires a point like s. More specifically, every interval that begins at or before p and extends to the right of p must contain $f(b)$, and every interval that ends at q or thereafter and extends to the left of q contains $f(a) + \rho(a)$.

Since $(X, \sim)$ is connected and $C \cup D \neq \emptyset$, there must be points like r and/or s, with both present if C and D are both nonempty. Let B be the set of points whose intervals lie strictly between p and q and do not intersect all such intervals. Then B is a buried subgraph whose K-set is the set of all points like r and s, plus points with intervals strictly between p and q that intersect all intervals for B. Note that B contains a and b with $a \not\sim b$, its K-set is nonempty and contains no point in B, and every path from a point in B to a point not in B uses a point in the K-set. □

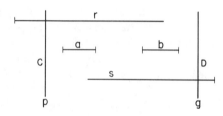

Figure 3.8

4

Betweenness

This chapter relates interval graphs, interval orders, and other binary relations to ternary betweenness relations. It shows what must be true of a betweenness relation T on X so that T induces in a natural way a specified type of binary relation on X. Main theorems characterize interval orders and semiorders that agree with T by simple conditions on betweenness. Other theorems identify conditions on T for the existence of an interval graph or indifference graph that corresponds in a natural way to a betweenness relation, and for the existence of a weak order or linear order that agrees with T.

4.1 BETWEENNESS RELATIONS

It is assumed throughout the chapter that X is a nonempty set and T is a subset of $X^3 = X \times X \times X$. Thus (X, T) is a relational system in which T is a ternary relation on X. For convenience we shall write an ordered triple (x, y, z) in X^3 as xyz and, when T is a betweenness relation and $xyz \in T$, we shall say that y is between x and z.

A ternary relation T is *symmetric* if, for all $x, y, z \in X$,

$$xyz \in T \Rightarrow zyx \in T.$$

We define T or (X, T) as a *betweenness relation* if T is symmetric. Thus, like the notion of a graph as a symmetric related set, the notion of a betweenness relation is very general.

As with graphs, there are nonstrict (reflexive) and strict (irreflexive) versions of betweenness. When $X \subseteq \mathbb{R}$, the usual nonstrict version is

$$xyz \in T \Leftrightarrow x \leq y \leq z \quad \text{or} \quad z \leq y \leq x,$$

and the usual strict version is

$$xyz \in T \Leftrightarrow x < y < z \quad \text{or} \quad z < y < x.$$

When $X \subseteq \mathbb{R}^2$, strict betweenness has several natural definitions that depend on the definition for $<$ in the preceding expression. One version of $<$ is the lexicographic linear order $(x_1, x_2) < (y_1, y_2) \Leftrightarrow [x_1 < y_1 \text{ or } (x_1 = y_1, x_2 < y_2)]$.

Another is $(x_1, x_2) < (y_1, y_2) \Leftrightarrow [x_1 < y_1 \text{ and } x_2 < y_2]$. A third is $(x_1, x_2) < (y_1, y_2) \Leftrightarrow [x_1 \leq y_1, x_2 \leq y_2, \text{ and } (x_1, x_2) \neq (y_1, y_2)]$. In the latter two cases, $(X, <)$ is a poset.

We define a betweenness relation T as *strict* if, for all $x, y, z \in X$,

$$xyz \in T \Rightarrow yxz \notin T.$$

In conjunction with symmetry, this implies that whenever y is between x and z it equals neither x nor z. However, it leaves open the possibility that $xyx \in T$. For example, if (X, I) is an irreflexive tree graph (no loops, no simple cycles of length three or more), and if $xyz \in T$ means that xIy and yIz, then (X, T) is a strict betweenness relation that has $xyx \in T$ if and only if xIy.

We also define a betweenness relation T as *strong* if, for all $x, y, z \in X$,

$$xyz \in T \Rightarrow x \neq y \neq z \neq x.$$

The tree relation just defined is strict but not generally strong; those noted earlier for $<$ on $\mathbb{R}^n$ are both strict and strong. Examples of strong betweenness relations that are not strict include the set of all triples in X^3 whose points are distinct ($|X| \geq 3$), and the set of all length-2 simple paths in an irreflexive graph that contains a triangle (simple 3-cycle).

In ensuing sections of the chapter we shall focus on strict betweenness relations that are also strong since we shall be concerned with betweenness relations that agree with partial orders of various types. In general, we shall say that a binary relation $\prec$ on X *agrees with* (X, T) if, for all $x, y, z \in X$,

$$xyz \in T \Leftrightarrow (x \prec y \prec z \text{ or } z \prec y \prec x),$$

where $x \prec y \prec z$ means that $x \prec y$, $y \prec z$, and $x \prec z$. The bulk of our proofs will consist of showing that certain conditions on a strict betweenness relation imply the existence of an interval order (semiorder, weak order, linear order) that agrees with the betweenness relation.

Two graphs, one the complement of the other, are defined for every betweenness relation (X, T) as follows. For all $x, y \in X$, let

$x \sim y$ if $x = y$ or x and y never appear in the same triple in T;

$x \not\sim y$ if $x \neq y$ and x and y both appear in at least one triple in T.

We refer to $(X, \sim)$ and $(X, \not\sim)$ as the *graphs on X induced by T*: $(X, \sim)$ is the *reflexive graph induced by T*, and $(X, \not\sim)$ is the *irreflexive graph induced by T*. Our main graph theorems show precisely what must be true of a strict betweenness relation so that $(X, \sim)$ is an interval graph or an indifference graph. *Within this chapter, $\sim$ and $\not\sim$ are used only in the ways they are defined in this paragraph.*

Interest in ternary betweenness relations stems from their use in axiomatizations of geometry in the latter part of the nineteenth century and early in this century (Pasch, 1882; Peano, 1889, 1894; Hilbert, 1899; Veblen, 1904). More recent treatments of geometry that include betweenness as an undefined primitive include Robinson (1940) and Blumenthal (1961).

Huntington and Kline (1917) identify 11 sets of axioms for T that are necessary and sufficient for the existence of an agreeing linear order, and Huntington (1924) added another set to these. All 12 sets use the strict and strong properties defined in the preceding paragraphs in addition to the completeness axiom

$$x \neq y \neq z \neq x \Rightarrow \{xyz, zxy, yzx\} \cap T \neq \emptyset.$$

Because this will be used later, we note here that T is said to be *complete* if it satisfies this condition for all $x, y, z \in X$.

The other axioms used by Huntington and Kline are transitivity conditions like

$$xab, aby \in T \Rightarrow xay \in T,$$
$$xab, ayb \in T \Rightarrow xay \in T.$$

These two in conjunction with strictness, strongness, and completeness (and of course symmetry) comprise one of the 11 Huntington-Kline sets.

Other axiom sets for reflexive ($\leq$) versions of linear orders and partial orders are noted by Altwegg (1950) and Sholander (1952). Sholander's axioms (on nonstrict T) for an agreeing reflexive partial order include a condition that looks very much like the comparability graph condition of Theorem 1.7 which says that every cycle of odd length has a triangular chord. Indeed, thanks to Theorem 1.7, we know that a strict betweenness relation admits an agreeing (irreflexive) partial order if and only if it is strong and its induced graph $(X, \nsim)$ is a comparability graph.

Additional background on betweenness, including its use in lattices, related notions of cyclic order, and specialized definitions arising from quaternary and binary relations, is given in Fishburn (1971). Specific motivation for the characterizations of interval graphs (Theorem 2) and indifference graphs (Theorem 4) in terms of T that are given later owes a great deal to Goodman's (1951) notion of local betweenness based on matching or similarity relations, and to Roberts's subsequent (1970, 1971) connection between betweenness and indifference graphs.

4.2 AXIOMS AND PRELIMINARY LEMMAS

The purpose of this section is to complete the list of axioms that are used to characterize betweenness relations that have agreeing interval orders and to note a few simple lemmas that follow from these axioms. The main theorem for interval orders and its correspondent for interval graphs are presented in the next section. Section 4.4 contains the theorems for semiorders and indifference graphs, and the final section comments briefly on conditions for agreeing weak orders and linear orders.

Recall that $x \not\leftrightarrow y$ if $x \neq y$ and x and y both appear in some triple in T. We shall use the following three axioms, which apply to all $x, y, z, a, b, c \in X$:

A1. $(xyz \in T, z \not\leftrightarrow y) \Rightarrow |\{xya\} \cap T| + |\{ayz\} \cap T| = 1$.
A2. $x \not\leftrightarrow y \not\leftrightarrow z \not\leftrightarrow x \Rightarrow \{xyz, zxy, yzx\} \cap T \neq \emptyset$.
A3. $abc, xyz \in T \Rightarrow \{abx, abz, xyc, zyc\} \cap T \neq \emptyset$.

These are easily visualized when $x, y, \ldots$ are thought of as points or intervals in a linearly ordered set. A1 says that if y is between x and z, and if a is distinct from y but is related to y in some betweenness triple, then either y is between x and a, or y is between a and z, *and not both*.

Axiom A2 is an obvious weakening of completeness. It says that if each pair of points in $\{x, y, z\}$ is in some betweenness triple, then one of the points must be between the other two. It is easily checked that A1 and A2 must hold if there is a partial order $\prec$ on X that agrees with (X, T). For example, the hypotheses of A2 would then imply

$$(x \prec y \text{ or } y \prec x) \quad \text{and} \quad (y \prec z \text{ or } z \prec y) \quad \text{and} \quad (x \prec z \text{ or } z \prec x),$$

and these yield $(x \prec y \prec z \text{ or } z \prec y \prec x)$ or $(z \prec x \prec y \text{ or } y \prec x \prec z)$ or $(y \prec z \prec x \text{ or } x \prec z \prec y)$, hence $xyz \in T$ or $zxy \in T$ or $yzx \in T$.

The same result does not hold for A3, which is designed explicitly for agreeing *interval* orders and need not hold for agreeing *partial* orders. If $\prec$ on X is an interval order that agrees with (X, T), then the hypotheses of A3 imply

$$(a \prec b \prec c \text{ or } c \prec b \prec a) \quad \text{and} \quad (x \prec y \prec z \text{ or } z \prec y \prec x).$$

Assume $a \prec b \prec c$ for definiteness. Then, using (2.1) of Section 2.1,

$$x \prec y \prec z \Rightarrow (b \prec z \text{ or } y \prec c) \Rightarrow (abz \in T \text{ or } xyc \in T),$$

$$z \prec y \prec x \Rightarrow (b \prec x \text{ or } y \prec c) \Rightarrow (abx \in T \text{ or } zyc \in T).$$

Hence A3 is necessary for an agreeing interval order.

Lemma 1. *If $(X, \prec)$ is an interval order that agrees with a betweenness relation (X, T), then T is strict, strong, and satisfies A1, A2, and A3.*

Our next lemma notes consequences of A1.

Axioms and Preliminary Lemmas

Lemma 2. *Suppose* (X, T) *is a strict betweenness relation that satisfies* A1. *Then*

C1. T *is strong,*
C2. $xyz \in T \Rightarrow \{yxz, yzx\} \cap T = \varnothing$,
C3. $xyz, xyw \in T \Rightarrow wyz \notin T$,
C4. $xyz, yzw \in T \Rightarrow \{xyw, xzw\} \subseteq T$,
C5. $xyz, xwy \in T \Rightarrow \{wyz, xwz\} \subseteq T$.

Proof. Strictness prohibits $xxz \in T$, and this and A1 prohibit $xyx \in T$ when $x \neq y$. C2 follows directly from symmetry and strictness. Given $xyz, xyw \in T$ for C3, A1 implies xyw or wyz, and not both. Since $xyw \in T$ by hypothesis, $wyz \notin T$.

A1 and the hypotheses of C4 imply $xyw \in T$ or $wyz \in T$, and not both. Since $wyz \notin T$ by C2 and symmetry, $xyw \in T$; $xzw \in T$ is proved similarly.

For C5, with $xyz \in T$ and $xwy \in T$, A1 implies $wyz \in T$ or $xyw \in T$. Since $xyw \notin T$ by C2, $wyz \in T$. The other part of the conclusion of C5 then follows from C4. □

Given a strict betweenness relation (X, T), define $x_1 x_2 \ldots x_n$ ($n \geq 3$) as a *T-string* if $x_i x_j x_k \in T$ for all $1 \leq i < j < k \leq n$. Consequences C4 and C5 of Lemma 2 show two ways of constructing strings. The addition of A2 in the next lemma allows additional constructions: C6 shows how an element x can be inserted in the middle of a string, and C7 gives another way of extending a string at one end.

Lemma 3. *Suppose* (X, T) *is a strict betweenness relation that satisfies* A1 *and* A2. *Then*

C6. $\{abc, bcd, abx, xcd\} \subseteq T \Rightarrow bxc \in T$,
C7. $(xab, yab \in T, x \neq y) \Rightarrow \{xya, yxa\} \cap T \neq \varnothing$.

Proof. Use A2 and then C3 in each case. □

The final lemma of this section presents an important consequence of A1–A3.

Lemma 4. *Suppose* (X, T) *is a strict betweenness relation that satisfies* A1, A2, *and* A3. *Then*

C8. $\{abp, xyp\} \subseteq T \Rightarrow \{ayp, xbp, bpy\} \cap T \neq \varnothing$.

Proof. Assume that abp and xyp are in T. By symmetry and A3, $\{pyx, pba\} \subseteq T \Rightarrow \{pyp, pya, pbx, abx\} \cap T \neq \varnothing$. Therefore, by symmetry and C1,

$$\{ayp, xbp, abx\} \cap T \neq \varnothing.$$

Since ayp and xbp appear in the conclusion of C8, suppose that $abx \in T$. Then, using C7, $x \neq p$ along with $abx, abp \in T$ implies either $bpx \in T$, which yields $ypb \in T$ with the use of $pyx \in T$ and C5, or $bxp \in T$. If $bxp \in T$ then this, along with $\{abp, xyp, abx\} \subseteq T$, yields the T-string $abxyp$ by applications of C4 and C5, so in particular $ayp \in T$. □

4.3 INTERVAL ORDERS AND GRAPHS

Theorem 1. *There is an interval order on X that agrees with a strict betweenness relation (X, T) if and only if (X, T) satisfies* A1, A2, *and* A3.

The necessity of the conditions on (X, T) in Theorem 1 for the existence of an agreeing interval order has already been noted in Lemma 1. The main task of this section will be to prove the sufficiency of those conditions for an agreeing interval order.

Before doing that, we note several facts about interval graphs in relation to Theorem 1. First, as is evident from Theorem 3.1, if $(X, \prec)$ is an interval order that agrees with (X, T), then $(X, sc(\prec))$ is an interval graph. Moreover, $sc(\prec) \subseteq \sim$, where $(X, \sim)$ is the reflexive graph induced by T. This follows from the fact that if not$(x \sim y)$, then $x \not\sim y$, so x and y are in some triple in T and therefore $x \prec y$ or $y \prec x$, hence not$(xsc(\prec)y)$. Thus, not$(x \sim y) \Rightarrow$ not$(xsc(\prec)y)$, so $xsc(\prec)y \Rightarrow x \sim y$.

However, $\sim$ need not be equal to $sc(\prec)$ when $(X, \prec)$ is an interval order that agrees with (X, T). One reason for this is that there can be a pair of points in no triple of T such that $\prec$ must hold between the two for every $\prec$ that agrees with (X, T). A case in point is illustrated in Fig. 4.1, where T has four triples and their symmetric duals. The interval order shown in the figure clearly agrees with (X, T). Moreover, *every* such agreeing interval order must have either $x \prec c$ or $c \prec x$, so that not$(xsc(\prec)c)$. However, $x \sim c$ since x and c appear together in no triple in T.

The preceding example suggests the following definition. A pair of points $\{x, c\}$ in X is a *weak pair* of (X, T) if $x \sim c$ (x and c not in the same triple of T, or $x = c$) and if there exist $a, b, y, z \in X$ such that

$$\{abc, xyz, ayz\} \subseteq T \quad \text{and} \quad byz \notin T.$$

The following theorem shows the usefulness of this definition.

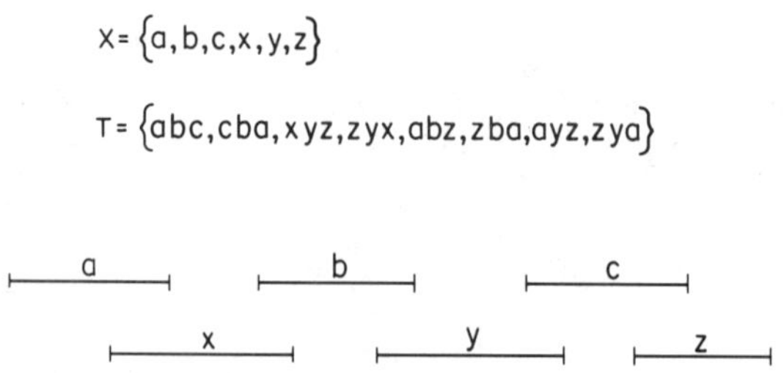

$$X = \{a, b, c, x, y, z\}$$

$$T = \{abc, cba, xyz, zyx, abz, zba, ayz, zya\}$$

Figure 4.1

Theorem 2. *The reflexive graph induced by a strict betweenness relation that satisfies A1, A2, and A3 is an interval graph if and only if the betweenness relation has no weak pair.*

We note also that the reflexive graph induced by a strict betweenness relation can be an interval graph even when the betweenness relation does *not* satisfy A1 through A3. Suppose, for example, that $X = \{a, b, x, y, z, w\}$ and that

$$T = \{xyz, xyw, xzw, yzw, wab, waz, wyb, \text{ and their symmetric duals}\}.$$

Then the pairs in $(X, \sim)$ are (x, a), (x, b), (y, a), (z, b), the duals of these, and all (t, t) for $t \in X$. As shown in Fig. 4.2, $(X, \sim)$ is an interval graph. Note, however, that every interval order $\prec$ that agrees with $(X, \sim)$ must have $y \prec x \prec z$ or $z \prec x \prec y$, whereas $xyz \in T$. Condition A1 fails for (X, T) since $xyz \in T$ and $b \not\sim y$, yet neither xyb nor byz is in T. Thus, although $(X, \sim)$ has an agreeing interval order (Fig. 4.2), (X, T) does not.

The proof of Theorem 2 appears at the end of this section. Since it relies on certain constructions and results established in the sufficiency proof of Theorem 1, we now turn to that proof.

Throughout the rest of this section, assume that (X, T) is a strict betweenness relation that satisfies A1–A3. For notational convenience, write $xyz \in T$ simply as xyz, omitting the membership designation, and let $\neg xyz$ stand for $xyz \notin T$. A T-string $x_1 x_2 \ldots x_n$ will also be written without qualification: see C4 and C5 below.

Since frequent use will be made of the axioms and the consequences in Lemmas 2 through 4, we list these here in the abbreviated notational format.

A1. $(xyz, a \not\sim y) \Rightarrow xya$ or ayz, and not both.
A2. $x \not\sim y \not\sim z \not\sim x \Rightarrow xyz$ or zxy or yzx.
A3. $(abc, xyz) \Rightarrow abx$ or abz or xyc or zyc.
C1. $xyz \Rightarrow x \neq y \neq z \neq x$.
C2. $xyz \Rightarrow \neg yxz$ and $\neg yzx$.
C3. $(xyz, xyw) \Rightarrow \neg wyz$.
C4. $(xyz, yzw) \Rightarrow xyzw$.
C5. $(xyz, xwy) \Rightarrow xwyz$.
C6. $(abc, bcd, abx, xcd) \Rightarrow bxc$.
C7. $(xab, yab, x \not\sim y) \Rightarrow xya$ or yxa.
C8. $(abp, xyp) \Rightarrow ayp$ or xbp or bpy.

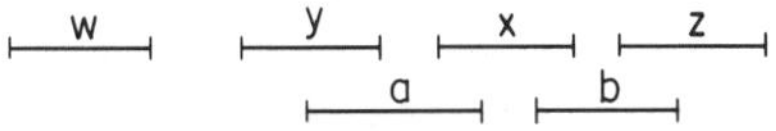

Figure 4.2

Sufficiency Proof of Theorem 1. If T is empty, the desired sufficiency conclusion of Theorem 1 follows immediately with $\prec = \emptyset$. Assume henceforth that T is not empty. Our aim is to construct an interval order on X that agrees with T. We shall do this by defining binary relations $<_1$ and $<_2$ on X and eventually show that the union of $<_1$ and $<_2$ is an interval order that agrees with T. Until later in the proof we shall work only with $<_1$.

Fix abc since T is not empty. (By notational convention, abc is in T.) Define $<_1$ in groupings of three as follows:

$$x <_1 y <_1 z \quad \text{if } xyz \text{ and } [(abz \text{ and } \neg bzy) \text{ or } (xyc \text{ and } \neg bcy)].$$

It is understood that $x <_1 y <_1 z$ means $x <_1 y$, $y <_1 z$, and $x <_1 z$. By symmetry and strictness, $a <_1 b <_1 c$. By C1, $<_1$ is irreflexive. We shall prove that $<_1$ is asymmetric, and then show that it agrees with T.

Lemma 5. $<_1$ *is asymmetric.*

Proof. Contrary to the lemma, suppose $x <_1 y$ and $y <_1 x$. For $x <_1 y$, the definition of $<_1$ implies that one of the following three things is true for some $t \in X$:

1. xyt and $[(abt, \neg bty)$ or $(xyc, \neg bcy)]$.
2. xty and $[(aby, \neg byt)$ or $(xtc, \neg bct)]$.
3. txy and $[(aby, \neg byx)$ or $(txc, \neg bcx)]$.

Similarly, for $y <_1 x$, one of the following holds for some $p \in X$:

4. yxp and $[(abp, \neg bpx)$ or $(yxc, \neg bcx)]$.
5. ypx and $[(abx, \neg bxp)$ or $(ypc, \neg bcp)]$.
6. pyx and $[(abx, \neg bxy)$ or $(pyc, \neg bcy)]$.

These yield nine major cases $(1 \& 4, 1 \& 5, \ldots, 3 \& 6)$ and four subcases (the "or" options) within each major case. We need to show that every combination yields a contradiction. The subcases that take the first options (ab_* in both instances) are easily disposed of. For example, for $1 \& 6$ suppose $(abt, \neg bty)$ and $(abx, \neg bxy)$. Then, since $x \not\sim t$ by 1, symmetry and C7 imply either bxt (hence $bxyt$ by C5, which contradicts $\neg bxy$) or btx (hence $btyx$ by C5, which contradicts $\neg bty$).

The symmetry within the nine main cases shows that it will suffice to consider only the following six combinations: $1 \& 4$, $1 \& 5$, $1 \& 6$, $2 \& 5$, $2 \& 6$ and $3 \& 6$. The subcases of these six that use the second options ($_{**}c$ in both instances) are dealt with as follows: $1 \& 4$ contradicts strictness; $1 \& 5$ gives $xypc$ by C5, which contradicts ypx in 5; $1 \& 6$ implies $\neg xyp$ by C3, contrary to pyx in 6; $2 \& 6$ implies tcy by C6 [since $xtyp$, xtc, cyp], so that, by A1, either tcb (contrary to $\neg bct$) or bcy (contrary to $\neg bcy$); $3 \& 6$ yields xcy by C6,

Interval Orders and Graphs

which by A1 implies either xcb (contrary to $\neg bcx$) or bcy (contrary to $\neg bcy$). This leaves 2 & 5, which for the subcase under consideration has

$$xty, xtc, \neg bct, ypx, ypc, \neg bcp.$$

By C8, $(xtc, ypc) \Rightarrow xpc$ or ytc or pct. First, xpc in conjunction with ypc implies $\neg ypx$ by C3, a contradiction. Second, ytc and ytx imply $\neg ctx$ by C3, another contradiction. Third, pct and $b \nrightarrow c$ imply by A1 that either bct (contrary to $\neg bct$) or pcb (contrary to $\neg bcp$).

This leaves twelve subcases (the mixed options) for the six major cases under consideration. We detail each of these, using notational similarity when possible.

1 & 4: xyt and yxp. Hence $pxyt$ by C4.

Mixed option (i): $(xyc, \neg bcy, abp, \neg bpx)$. Then $pxyc$ by C4, so pyc. Since pba also, C8 gives either pbc (which with pba implies $\neg abc$ by C3, a contradiction) or bpy (hence $bpxy$ from C5, hence bpx, a contradiction) or pya. By C7, pya, pyc, and $a \nrightarrow c$ imply yca (hence $ycba$ by C5, then ycb, a contradiction) or yac (hence $pyabc$, so pab, a contradiction).

Mixed option (ii): $(abt, \neg bty, yxc, \neg bcx)$. Similar to mixed option (i).

1 & 5: xyt and ypx. Hence $xpyt$ by C5.

(i): $(abt, \neg bty, ypc, \neg bcp)$. Interchange of x and p gives similarity to 1 & 4(ii).

(ii): $(xyc, \neg bcy, abx, \neg bxp)$. Interchange x and p, note that pyc and pxy imply xyc by C5, and use 1 & 4(i).

1 & 6: xyt and xyp.

(i): $(abt, \neg bty, pyc, \neg bcy)$. By C8, abt and xyt imply either bty (a contradiction) or ayt or xbt.

Consider ayt first. This gives $a \nrightarrow y$, so by A1 with xyp, either xya (which by C3 with tya gives $\neg xyt$, a contradiction) or ayp. Then, by C7, ayp, cyp and $a \nrightarrow c$ imply either acy (hence $abcyp$, so bcy, a contradiction) or cay (hence $cbayt$, so bat, a contradiction under C2).

To complete subcase (i), suppose xbt. Then, by A1, abc and $x \nrightarrow b$ yield either abx (which with tbx implies $\neg abt$, a contradiction) or xbc. Hence xbc. By C8 for xbc and xyp, get xyc (which with pyc implies $\neg xyp$, a contradiction) or bxy (which yields $bxyt$, hence bxt, a contradiction to xbt) or xbp. Hence xbp. Since $a \nrightarrow b$, A1 implies either xba (a contradiction to xbc and abc under C3) or abp. Hence apb. By C8 on apb and cyp, either cbp (a contradiction with abp and abc) or bpy (hence $xbpy$, so xpy, a contradiction to xyp) or ayp. Hence ayp. By C7, ayp, cyp and $a \nrightarrow c$ imply either acy (hence $abcyp$, so bcy, a contradiction to $\neg bcy$) or cay (hence $cbayp$, so bap, a contradiction to abp).

(ii): $(xyc, \neg bcy, abx, \neg bxy)$. C8 on abx and xyc gives cbx (a contradiction to abx and abc under C3) or bxy (contrary to $\neg bxy$) or ayx. Then C7 applied to *ayx* and *xyc* gives *acy* (hence *abcyx*, a contradiction to $\neg bcy$) or *cay* (hence *cbayx*, a contradiction to *abx*).

2 & 5: *xty* and *xpy*.
(i): $(aby, \neg byt, ypc, \neg bcp)$. C8 on *aby* and *cpy* implies either *apy* [which with *cpy* and C7 implies either *acp* (hence *abcpy*, contrary to $\neg bcp$) or *cap* (hence *cbapy*, contrary to *aby*)] or *cby* (which contradicts *aby* and *abc* under C3) or *byp* (hence *bypx*, so *byx*, then *bytx* by C5, a contradiction to $\neg byt$).
(ii): $(xtc, \neg bct, abx, \neg bxp)$. Similar to (i).

2 & 6: *xty* and *pyx*. Then *xtyp* by C5.
(i): $(aby, \neg byt, pyc, \neg bcy)$. By A1 for *pyc* and $b \not\prec y$, either *byc* [which with $y \not\prec t$ under A1 implies either *byt* (contrary to $\neg byt$) or *tyc* (a contradiction to *pyc* and *pyt* under C3)] or *pyb* [which with *pyc* under C7 gives either *ybc* (contrary to *yba* and *abc*) or *ycb* (hence *pycba*, contrary to $\neg bcy$)].
(ii): $(xtc, \neg bct, abx, \neg bxy)$. By C8, *abx* and *ctx* imply either *atx* [which with *ctx* under C7 implies either *act* (hence *abct*, contrary to $\neg bct$) or *cat* (hence *cbatx*, a contradiction to *abx*)] or *cbx* (a contradiction to *abx* and *abc* under C3) or *bxt* (hence *bxty*, contrary to $\neg bxy$).

3 & 6: *txy* and *pxy*. Hence *txyp* by C4.
(i): $(aby, \neg byx, pyc, \neg bcy)$. Like 2 & 6(i) with *x* and *t* interchanged.
(ii): $(txc, \neg bcx, abx, \neg bxy)$. Similar to (i). □

Lemma 6. $xyz \Leftrightarrow (x <_1 y <_1 z \text{ or } z <_1 y <_1 x)$.

Proof. Assume *xyz* first. By the definition of $<_1$, we wish to show that *abc* and *xyz* imply one of $(abz, \neg bzy)$, $(xyc, \neg bcy)$, $(abx, \neg bxy)$, and $(zyc, \neg bcy)$ since the first two of these give $x <_1 y <_1 z$ and the second two give $z <_1 y <_1 x$. The use of A3 on (abc, xyz) yields *abz* or *abx* or *xyc* or *zyc*. Suppose for definiteness that *abz* since the other cases are similar by symmetry. Given *abz*, if $\neg bzy$ then we are done, so assume *bzy*. Then, with C4, *abzyx*, so *abx* and *byx*. Since *byx* implies $\neg bxy$ by C2, we get $(abx, \neg bxy)$.

To prove the converse, suppose $x <_1 y <_1 z$ or $z <_1 y <_1 x$. Then, by A2 since $x \not\prec y \not\prec z \not\prec x$, either *xyz* or *zxy* or *yzx*. If *zxy* then, by the preceding paragraph, either $z <_1 x <_1 y$ or $y <_1 x <_1 z$, each of which contradicts asymmetry as proved in Lemma 5 regardless of which of $x <_1 y <_1 z$ and $z <_1 y <_1 x$ holds. A similar contradiction obtains if *yzx*. Hence *xyz*. □

We prove one more lemma for $<_1$ before $<_2$ is introduced.

Lemma 7. $(x <_1 y <_1 z, p <_1 q <_1 r) \Rightarrow [(p <_1 q <_1 z \text{ or } x <_1 y <_1 r) \text{ and } (p <_1 y <_1 z \text{ or } x <_1 q <_1 r)]$.

Proof. Assume that $x <_1 y <_1 z$ and $p <_1 q <_1 r$. By A3, *pqx* or *pqz* or *xyr* or *zyr*. By Lemmas 5 and 6, $pqz \Rightarrow p <_1 q <_1 z$, and $xyr \Rightarrow x <_1 y <_1 r$.

Interval Orders and Graphs 67

Suppose instead that pqx. Then, by Lemmas 5 and 6, $p <_1 q <_1 x$. By C8, pqx and xyz imply zqx (which by Lemmas 5 and 6 requires $x <_1 q <_1 z$ since $x <_1 z$; but then $x <_1 q$, a contradiction to asymmetry since $p <_1 q <_1 x$) or pyx (which yields a similar contradiction) or qxy. Hence

$$pqx \Rightarrow qxy \Rightarrow pqxyz \Rightarrow pqz \Rightarrow p <_1 q <_1 z.$$

In like manner, $zyr \Rightarrow qry \Rightarrow pqryz \Rightarrow pqz \Rightarrow p <_1 q <_1 z$. Therefore the hypotheses of Lemma 7 imply ($p <_1 q <_1 z$ or $x <_1 y <_1 r$).

The proof of ($p <_1 y <_1 z$ or $x <_1 q <_1 r$) is similar. □

The example of Fig. 4.1 with $X = \{x, y, z, a, b, c\}$ and

$$T = \{abc, cba, xyz, zyx, abz, zba, ayz, zya\}$$

shows that $<_1$ need not be an interval order since $b <_1 c$ and $x <_1 y$, but neither $b <_1 y$ nor $x <_1 c$. We therefore define another binary relation $<_2$ on X as follows:

$$x <_2 y \quad \text{if } (x <_1 r <_1 s, p <_1 q <_1 y, \neg qrs) \quad \text{for some } p, q, r, s.$$

In addition, let

$$\prec = <_1 \cup <_2.$$

To show that $\prec$ is an interval order, we note first that it satisfies (2.1).

Lemma 8. $(x \prec y, z \prec w) \Rightarrow (x \prec w \text{ or } z \prec y)$.

Proof. Assume $x \prec y$ and $z \prec w$. Suppose first that $x <_1 y$ and $z <_1 w$. Then, when one of $x <_1 y <_1 t$, $x <_1 t <_1 y$, and $t <_1 x <_1 y$ is paired with one of $z <_1 w <_1 v$, $z <_1 v <_1 w$, and $v <_1 z <_1 w$, Lemma 7 implies immediately that $x <_1 w$ or $z <_1 y$ except in the following two cases:

(i) $x <_1 y <_1 t$ and $v <_1 z <_1 w$. If zyt then $z <_1 y$. If $\neg zyt$ then $x <_2 w$ by the definition of $<_2$. Hence either $z \prec y$ or $x \prec w$.

(ii) $z <_1 w <_1 v$ and $t <_1 x <_1 y$. Similar to (i).

Suppose next that $x <_2 y$ and $z <_2 w$. For $x <_2 y$, use the preceding definition of $<_2$ as stated, and for $z <_2 w$ use

$$(z <_1 j <_1 k, m <_1 n <_1 w, \neg njk).$$

By Lemma 7, either $x <_1 j <_1 k$ (whence $x <_2 w$) or $z <_1 r <_1 s$ (whence $z <_2 y$).

Finally, suppose that $x <_2 y$ and $z <_1 w$. Use the defining statement for $x <_2 y$. If $z <_1 w <_1 v$ then, by Lemma 7, either $x <_1 w <_1 v$ (so $x \prec w$) or $z <_1 r <_1 s$ (so $z <_2 y$, hence $z \prec y$). If $z <_1 v <_1 w$, Lemma 7 implies $x <_1 v$

$<_1 w$ (so $x \prec w$) or $z <_1 r <_1 s$ (so $z <_2 y$). If $v <_1 z <_1 w$, Lemma 7 implies $v <_1 z <_1 y$ (hence $z \prec y$) or $p <_1 q <_1 w$ (hence $x <_2 w$ and $x \prec w$). □

Two more easy steps show that $\prec$ is an interval order.

Lemma 9. $(p <_1 q <_1 x, x <_1 r <_1 s) \Rightarrow q <_1 x <_1 r$.

Proof. By C8 on srx and pqx, either sqx (hence $s <_1 q <_1 x$ by Lemmas 5 and 6 and $q <_1 x$; but then $s <_1 x$, a contradiction to $x <_1 r <_1 s$) or prx (also impossible by Lemmas 5 and 6) or qxr. Therefore qxr and, since $q <_1 x$ and $x <_1 r$, we have $q <_1 x <_1 r$. □

Lemma 10. $\prec$ *is an interval order*.

Proof. In view of C1 and Lemma 8, we need only show that $<_2$ is irreflexive. If $x <_2 x$, then $x <_1 r <_1 s$ and $p <_1 q <_1 x$ and $\neg qrs$ for the definition of $<_2$. But, by Lemma 9, $x <_1 r <_1 s$ and $p <_1 q <_1 x$ imply $q <_1 x <_1 r$, hence qxr, then $qxrs$ by C4, so qrs, contrary to $\neg qrs$. Therefore $x <_2 x$ is impossible. □

Thus, we have proved that $\prec$ on X is an interval order and $<_1$ satisfies

$$xyz \Leftrightarrow (x <_1 y <_1 z \text{ or } z <_1 y <_1 x).$$

To show that $\prec$ agrees with T, we need

$$xyz \Leftrightarrow (x \prec y \prec z \text{ or } z \prec y \prec x).$$

Since $x <_1 y <_1 z \Rightarrow x \prec y \prec z$, it remains only to prove

Lemma 11. $x \prec y \prec z \Rightarrow x <_1 y <_1 z$.

Proof. Assume that $x \prec y \prec z$. Suppose first that $x <_2 y$ and $y <_2 z$ with

$$x <_1 r <_1 s, \qquad p <_1 q <_1 y, \qquad \neg qrs,$$

$$y <_1 j <_1 k, \qquad m <_1 n <_1 z, \qquad \neg njk.$$

By Lemma 9, $q <_1 y <_1 j$. Apply Lemma 7 to this and $x <_1 r <_1 s$ to get either $q <_1 r <_1 s$ (contrary to $\neg qrs$) or $x <_1 y <_1 j$. Hence $x <_1 y <_1 j$. Also, by Lemma 7 on $q <_1 y <_1 j$ and $m <_1 n <_1 z$, either $q <_1 y <_1 z$ or $m <_1 n <_1 j$. [If $m <_1 n <_1 j$ then Lemma 7 with $y <_1 j <_1 k$ implies either $y <_1 j <_1 j$ (false) or $m <_1 n <_1 k$. Therefore mnk, which with mnj and $j \not\prec k$ implies (nkj or njk) by C7. If nkj then $j <_1 k <_1 n$ since $j <_1 k$, and thus $j <_1 n$, a contradiction to $m <_1 n <_1 j$. Hence njk. But this contradicts $\neg njk$ from the hypotheses.] Since $m <_1 n <_1 j$ yields a contradiction, $q <_1 y <_1 z$. Since $x \not\prec y$ by $x <_1 y <_1 j$, A1 implies xyz or qyx (a contradiction to qyj and xyj). Therefore xyz, and also $x <_1 y <_1 z$ by Lemma 6 since $x <_1 y$.

Suppose next that $x <_1 y$ and $y <_2 z$. Use $(y <_1 j <_1 k, m <_1 n <_1 z, \neg njk)$ for $y <_2 z$, and combine this with one of $x <_1 y <_1 t$, $x <_1 t <_1 y$, and $t <_1 x <_1 y$. If $t <_1 x <_1 y$, then Lemma 9 gives $x <_1 y <_1 j$. This and $m <_1 n <_1 z$ in Lemma 7 give $x <_1 y <_1 z$ or $m <_1 n <_1 j$, the latter of which is ruled out by the bracketed analysis in the preceding paragraph, so that $x <_1 y <_1 z$ as desired. If $x <_1 t <_1 y$, beginning with Lemma 9 we get $t <_1 y <_1 j \Rightarrow tyj \Rightarrow xtyj \Rightarrow xyj \Rightarrow x <_1 y <_1 j$, and the preceding sentence applies. If $x <_1 y <_1 t$, then Lemma 7 and $y <_1 j <_1 k$ imply $y <_1 y <_1 t$ (false) or $x <_1 j <_1 k$. By C7, xjk and yjk imply either xyj (hence $x <_1 y <_1 j$ and the first use of Lemma 7 in this paragraph applies) or yxj. If yxj, then $y <_1 x <_1 j$ since $x <_1 j$, but then $y <_1 x$ contradicts $x <_1 y <_1 t$. Hence $x <_1 y <_1 z$ in all cases.

Finally, suppose that $x <_2 y$ and $y <_1 z$. Use $(x <_1 r <_1 s, p <_1 q <_1 y, \neg qrs)$ for $x <_2 y$. We combine this with $y <_1 v <_1 z$, $y <_1 z <_1 v$, and $v <_1 y <_1 z$ in turn. If $y <_1 v <_1 z$, then Lemma 9 implies $q <_1 y <_1 v$, and Lemma 7 on this and $x <_1 r <_1 s$ implies either $q <_1 r <_1 s$ (contrary to $\neg qrs$) or $x <_1 y <_1 v$. Hence xyv, so $xyvz$, then xyz, and therefore $x <_1 y <_1 z$. If $y <_1 z <_1 v$ then $q <_1 y <_1 z$ by Lemma 9, hence $qyzv$ and $q <_1 y <_1 v$, and the two preceding sentences apply with $xyvz$ replaced by $xyzv$. Finally, if $v <_1 y <_1 z$, then Lemma 7 on this and $p <_1 q <_1 y$ implies $p <_1 q <_1 z$. Then C7 on $(pqz, pqy, y \not\sim z)$ gives qzy or qyz. If qzy then $y <_1 z <_1 q$ since $y <_1 z$; but then $p <_1 q <_1 y$ is contradicted. Therefore qyz and hence $q <_1 y <_1 z$. Lemma 7 with $x <_1 r <_1 s$ then yields $x <_1 y <_1 z$ since the other possibility $(q <_1 r <_1 s)$ contradicts $\neg qrs$. □

This completes the sufficiency proof of Theorem 1. We conclude this section with the proof of Theorem 2, using the same notations and definitions employed above.

Proof of Theorem 2. Let $(X, \sim)$ be the reflexive graph induced by a strict betweenness relation (X, T) that satisfies A1–A3, and assume that $T \neq \emptyset$ since otherwise there is nothing to prove. It follows easily from Lemma 6 that

$$\sim \; = sc(<_1).$$

Suppose first that (X, T) has no weak pair. We show that this implies that $<_2 \setminus <_1$ is empty. To the contrary, suppose $<_2 \setminus <_1$ is not empty, say

$$x <_2 c, \text{not}(x <_1 c), \text{ and therefore } x \sim c.$$

Since $x <_2 c$, there are $a, b, y, z \in X$ such that

$$x <_1 y <_1 z, \quad a <_1 b <_1 c, \quad \text{and} \quad \neg byz.$$

By Lemma 7, either $a <_1 b <_1 c$, which contradicts $x \sim c$, or $a <_1 y <_1 z$. Therefore $a <_1 y <_1 z$. But then

$$x \sim c, \quad \{abc, xyz, ayz\} \subseteq T, \quad \text{and} \quad byz \notin T,$$

so that $\{x,c\}$ is a weak pair, contrary to hypothesis. Hence $<_2 \setminus <_1 = \emptyset$ as claimed. Hence, by Lemma 10, $<_1$ is an interval order, and therefore $(X, \sim)$ is an interval graph.

Suppose next that $(X, \sim)$ is an interval graph. Then $<_1$ is an interval order and, by Lemma 6, $<_1$ is an interval order that agrees with (X, T). If (X, T) has a weak pair, say $\{x, c\}$ as in the final displayed expression in the preceding paragraph, then $b <_1 c$ and $x <_1 y$, but neither $x <_1 c$ (since $x \sim c$) nor $b <_1 y$ (since otherwise we get $b <_1 y <_1 z$, contrary to $\neg byz$), so $<_1$ is not an interval order, a contradiction. Therefore (X, T) has no weak pair. □

4.4 SEMIORDERS AND INDIFFERENCE GRAPHS

The following new axioms, which apply to all $a, b, c, d, x, y, z \in X$, will be used for agreeing semiorders:

A4. $abc, xyz \in T \Rightarrow \{abx, abz, xbc, zbc\} \cap T \neq \emptyset$.

A5. $abc, bcd \in T \Rightarrow \{abx, xcd, axd\} \cap T \neq \emptyset$.

Axiom A3 has the same hypotheses as A4 but a different conclusion, namely $\{abx, abz, xyc, zyc\} \cap T \neq \emptyset$. The distinctive feature of A4 is the use of two of a, b, and c in every triple in its conclusion along with the omission of y there. Indeed, if $a \prec b \prec c$ and either $x \prec y \prec z$ or $z \prec y \prec x$ when $\prec$ is a semiorder, then either $a \prec b \prec x$ or $a \prec b \prec z$ or $x \prec b \prec c$ or $z \prec b \prec c$, so A4 is necessary for the existence of an agreeing semiorder.

The hypotheses of A5 under conclusion C4 of Lemma 2 imply that $abcd$ is a T-string, with either $a \prec b \prec c \prec d$ or $d \prec c \prec b \prec a$ for an agreeing semiorder. If x is any other point in X, then (2.2) of Section 2.1 forces one of abx, xcd, and axd to be in T. For example, if $a \prec b \prec c \prec d$, then (2.2) implies ($a \prec x$ or $x \prec c$) and ($b \prec x$ or $x \prec d$), hence either $x \prec c \prec d$, $a \prec b \prec x$, or $a \prec x \prec d$.

Theorem 3. *There is a semiorder on X that agrees with a strict betweenness relation (X, T) if and only if (X, T) satisfies* A1 *through* A5.

Our preceding remarks along with Theorem 1 show that the conditions on (X, T) in Theorem 3 are necessary for the existence of an agreeing semiorder. The sufficiency proof of Theorem 3 is given after we discuss the relationship of Theorem 3 to indifference graphs.

Suppose (X, T) is a strict betweenness relation that satisfies A1–A5. There are three ways that can prevent $(X, \sim)$, the reflexive graph induced by (X, T), from being an indifference graph when T is not empty. First, (X, T) could have a weak pair (Fig. 4.1). Second, (X, T) could have no four-point T-string and have a point x in the relation $\sim$ to all three points in $abc \in T$. This is pictured in the left part of Fig. 4.3, where three placements are shown for x's interval. The first placement of x applies to $(X, \sim)$: since x intersects every

Semiorders and Indifference Graphs

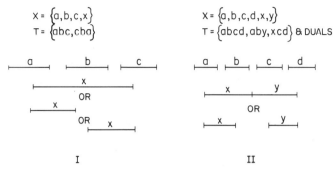

Figure 4.3

interval in $\{a, b, c\}$, $(X, \sim)$ is not an indifference graph. The second and third placements of x yield indifference graphs that have agreeing semiorders that also agree with (X, T). We shall construct such semiorders in Part I of the ensuing sufficiency proof of Theorem 3 through the introduction of an auxiliary binary relation $<_3$ that rectifies the situation illustrated on the left of Fig. 4.3.

The third way that $(X, \sim)$ can fail to be an indifference graph occurs when (X, T) has a four-point T-string, say $abcd$, but no five-point string, and there are points x and y for which

$$\text{not}(a \prec x), \quad x \prec c, \quad b \prec y, \quad \text{not}(y \prec d).$$

This is shown on the right of Fig. 4.3 with two placements for x and y. The upper placement adheres to $(X, \sim)$, with $x \sim y$ since x and y are not together in some triple in T. The lower placement gives an indifference graph with an agreeing semiorder that also agrees with (X, T). We construct such semiorders in Part II of the ensuing proof of Theorem 3 by introducing another auxiliary binary relation $<_4$.

A theorem for indifference graphs that parallels Theorem 2 for interval graphs would exclude the special cases described in Figs. 4.1 and 4.3. However, those cases become academic if there is a five-point T-string, so we state a simpler theorem for this case.

Theorem 4. *The reflexive graph induced by a strict betweenness relation that satisfies A1 through A5 and has a five-point T-string is an indifference graph.*

Proof. Given the hypotheses of the theorem for (X, T), it follows from the sufficiency proofs of Theorems 1 and 3 (the latter follows this proof) that there is a semiorder $\prec$ on X that agrees with (X, T) and includes $<_1$ (defined prior to Lemma 5). By agreement and Lemma 6,

$$(x \prec y \prec z \text{ or } z \prec y \prec x) \Leftrightarrow xyz \in T \Leftrightarrow (x <_1 y <_1 z \text{ or } z <_1 y <_1 x).$$

Because of the existence of a five-point T-string, a uniform representation for semiorders as in Theorem 2.7 shows that the special cases of $\sim$ associated with the situations illustrated in Fig. 4.1 ($x \sim c$) and Fig. 4.3 ($x \sim \{a,b,c\}$ for I, $x \sim y$ for II) cannot occur. In other words, the two points at issue (x and c; x and a or x and c; x and y) must be in some triple under $\prec$, hence in some triple in T, when the other conditions that describe those cases obtain. Thus every instance of $x \prec y$ corresponds to some triple in T that contains both x and y, so $\sim$ is in fact the symmetric complement of $\prec$, and $(X, \sim)$ is therefore an indifference graph. □

Sufficiency Proof of Theorem 3. Assume that (X, T) is a strict betweenness relation that satisfies A1–A5 and (for nontriviality) has $T \neq \emptyset$. Assume the notational conventions used in the sufficiency proof of Theorem 1 ($xyz, \neg xyz, x_1 x_2 \ldots x_n$) as well as the results of that proof. We divide the present proof into two parts according to whether T has a string of four points.

PART I. T has no four-point string. Thus, for all distinct x, y, z, and w in X, not(xyz and yzw). Fix abc as in the sufficiency proof of Theorem 1, define $\prec_1$, $\prec_2$ and $\prec = \prec_1 \cup \prec_2$ as in that proof, and define another binary relation $\prec_3$ on X by

$$x \prec_3 y \quad \text{if } x \prec p \prec q \text{ for some } p, q \in X, \text{ and } y \prec t \text{ for no } t \in X.$$

In addition, let $\prec^* = \prec \cup \prec_3$.

Lemma 12. $(x \prec^* y, y \prec^* z) \Rightarrow x \prec y \prec z$.

Proof. Since $x \prec_3 y$ implies not($y \prec^* z$), take $x \prec y$ for $x \prec^* y$. If $y \prec z$ then the conclusion of Lemma 12 follows from Theorem 1 (or Lemma 10). If $y \prec_3 z$, then $x \prec y$ and $y \prec r \prec s$ for some r, s. But then $xyrs$ is a four-point string, contrary to the hypothesis of Part I. □

Lemma 13. $\prec^*$ *is an interval order*.

Proof. Since $\prec$ and $\prec_3$ are irreflexive, so is $\prec^*$. To show that $(x \prec^* y, z \prec^* w) \Rightarrow (x \prec^* w \text{ or } z \prec^* y)$, consider cases:

1. $x \prec y$ and $z \prec w$. Use Lemma 10.
2. $x \prec_3 y$ and $z \prec_3 w$. Then $x \prec_3 w$ and $y \prec_3 z$ by the definition of $\prec_3$.
3. $x \prec_3 y$ and $z \prec w$. Take $x \prec p \prec q$ for $x \prec_3 y$. Then, by Lemma 8, either $x \prec w$ (hence $x \prec^* w$) or $z \prec p$ (hence $z \prec p \prec q$ and therefore $z \prec_3 y$).

Lemma 14. $(x \prec^* y, y \prec^* z) \Rightarrow (x \prec^* w \text{ or } w \prec^* z)$.

Proof. By Lemma 12, $(x \prec^* y, y \prec^* z) \to x \prec y \prec z$. If $w \prec t$ for no t, then $x <_3 w$, so $x \prec^* w$. Assume henceforth that $w \prec t$ for some t. Then, by the definitions of $<_1$ and $<_2$ and no four-point string, either $w <_1 p <_1 q$ or $p <_1 w <_1 q$ for some $p, q \in X$.

Suppose first that $w <_1 p <_1 q$ along with $x <_1 y <_1 z$ by Lemma 11. If $y <_1 p <_1 q$ also, then $xypq$ is a four-point T-string, a contradiction, and therefore $\neg ypq$. Hence $w <_2 z$ by the definition of $<_2$, and therefore $w \prec^* z$.

Suppose next that $p <_1 w <_1 q$. Then, by A4 applied to pwq and xyz, either $p \prec w \prec x$ (hence $p \prec w \prec x \prec y \prec z$, a contradiction) or $p \prec w \prec z$ (hence $w \prec^* z$) or $x \prec w \prec q$ (hence $x \prec^* w$) or $z \prec w \prec q$ (hence $x \prec y \prec z \prec w \prec q$, a contradiction). □

Since Lemmas 12, 13, and 14 in conjunction with Theorem 1 show that $\prec^*$ is a semiorder that agrees with (X, T), the proof of Part I is complete.

PART II. T has a four-point string, say $abcd$. Let $\prec$ be any interval order on X that agrees with (X, T), as guaranteed by Theorem 1. With $abcd$ fixed and $\prec$ fixed with $a \prec b \prec c \prec d$, define a binary relation $<_4$ on X by

$$x <_4 y \text{ if } (\text{not}(a \prec x), x \prec c, b \prec y, \text{not}(y \prec d)),$$

and let $<^* = \prec \cup <_4$.

Lemma 15. $(x <^* y, y <^* z) \Rightarrow x \prec y \prec z$.

Proof. Since $\prec$ is transitive, $x <_4 y \Rightarrow b \prec y \Rightarrow a \prec y \Rightarrow \text{not}(y <_4 z)$. Hence, both $<^*$ in the hypotheses of the lemma can't be $<_4$. If $x <_4 y$ and $y \prec z$ then $a \prec b \prec y \prec z$, and by A5 either $a \prec b \prec x$ or $a \prec x \prec z$ (each of which contradicts not$(a \prec x)$ from $x <_4 y$) or $x \prec y \prec z$. Hence $x \prec y \prec z$. Similarly, $(x \prec y, y <_4 z) \Rightarrow x \prec y \prec z$. □

It is easily seen that $<^*$ is irreflexive. By Theorem 1 and Lemma 15, $<^*$ agrees with (X, T). The next two lemmas complete the proof of Part II and hence of Theorem 3 by showing that $<^*$ is a semiorder.

Lemma 16. $(x <^* y, y <^* z) \Rightarrow (x <^* w \text{ or } w <^* z)$.

Proof. Assume $x <^* y$ and $y <^* z$. By Lemma 15, $x \prec y \prec z$. Using $a \prec b \prec c \prec d$ with w, A5 implies $a \prec b \prec w$ or $a \prec w \prec d$ or $w \prec c \prec d$. We consider these in turn.

1. $a \prec b \prec w$. This case is similar to that for $w \prec c \prec d$.

2. $a \prec w \prec d$. A4 on $a \prec w \prec d$ and $x \prec y \prec z$ implies either $a \prec w \prec x$ (hence $w \prec z$ by transitivity) or $a \prec w \prec z$ ($w \prec z$) or $x \prec w \prec d$ ($x \prec w$) or $z \prec w \prec d$ (hence $x \prec w$ by transitivity).

3. $w \prec c \prec d$. By A5 with z, either $a \prec b \prec z$ or $a \prec z \prec d$ or $z \prec c \prec d$. With either of the latter two we get $x \prec y \prec z \prec d$ and, using A5 on this with w, we get $x \prec w$ or $w \prec z$. This leaves $a \prec b \prec z$. If $a \prec w$ then A5 with z applied to $a \prec w \prec c \prec d$ implies either $a \prec w \prec z$ ($w \prec z$) or $z \prec c \prec d$ (so $x \prec y \prec z \prec d$ as before) or $a \prec z \prec d$ (so again $x \prec y \prec z \prec d$). On the other hand, if not($a \prec w$), then either $z \prec d$ (whence $x \prec y \prec z \prec d$ as before) or not($z \prec d$), in which case (not($a \prec w$), $w \prec c, b \prec z$, not($z \prec d$)) $\Rightarrow w <_4 z$. □

Lemma 17. $(x <^* y, z <^* w) \Rightarrow (x <^* w \text{ or } z <^* y)$.

Proof. Assume $x <^* y$ and $z <^* w$. If $x \prec y$ and $z \prec w$, the desired result is immediate since $\prec$ is an interval order. If $x <_4 y$ and $z <_4 w$, then $x <_4 w$ and $z <_4 y$ by the definition of $<_4$. Finally, suppose $x <_4 y$ and $z \prec w$. By the interval order property for $\prec$, $(x \prec c, z \prec w) \Rightarrow (x \prec w$ or $z \prec c)$. Suppose $z \prec c$ since $x \prec w$ is a desired conclusion. If not($a \prec z$) then $z <_4 y$, and if $a \prec z$ then $a \prec z \prec w$ and, by Lemma 16, either $x <^* w$ or $a <^* x$ (which is false since $a \prec x \Rightarrow$ not($x <_4 y$), and $a <_4 x \Rightarrow b \prec x \Rightarrow a \prec x$). Hence $x <_4 y$ and $z \prec w$ imply $x <^* w$ or $z <_4 y$. □

4.5 WEAK AND LINEAR ORDERS

We complete our discussion of betweenness by noting the changes in preceding axioms that lead to agreeing weak orders and linear orders. The only new assumption that we shall need is

A6. $(xyz \in T, a \in X) \Rightarrow \{ayz, xaz, xya\} \cap T \neq \emptyset$.

This is clearly necessary for an agreeing weak order.

Theorem 5. *There is a weak order on X that agrees with a strict betweenness relation (X, T) if and only if (X, T) satisfies A1, A2, and A6.*

Theorem 6. *There is a linear order on X that agrees with a strict betweenness relation (X, T) if and only if (X, T) satisfies A1 and is complete.*

Our sufficiency proof of Theorem 5 begins with a lemma which will allow us to make use of Theorem 1.

Lemma 18. *If (X, T) satisfies the conditions of Theorem 5, then it also satisfies A3.*

Proof. Assume that $abc, xyz \in T$, as in the hypotheses of A3. By A6 with c on xyz, either cyz or xcz or xyc. (As before, xyz denotes $xyz \in T$.) The conclusion of A3 is (abx or abz or xyc or zyc). Since cyz and xyc are in this conclusion, suppose xcz. Then xcz and $b \not\sim c$ under A1 imply either xcb (hence $xcba$, then xba) or bcz (hence $abcz$, so abz). Therefore $xcz \Rightarrow (abx$ or $abz)$. □

Given the conditions on (X, T) in Theorem 5 and the result of Lemma 18, let $\prec$ be the interval order on X constructed in the sufficiency proof of Theorem 1 ($\prec = <_1 \cup <_2$) and let $I = sc(\prec)$. Since $\prec$ is a weak order when I is transitive, the sufficiency proof of Theorem 5 will be completed by proving that I is transitive. This is obvious if $T = \emptyset$, so assume that T is not empty. We note

Lemma 19. $x \not\sim y \Leftrightarrow (x \prec y$ or $y \prec x)$.

Proof. Clearly, $x \not\sim y \Rightarrow (x \prec y$ or $y \prec x)$. Conversely, suppose $x \prec y$. If $x <_1 y$ then $x \not\sim y$. If $x <_2 y$ take

$$x <_1 r <_1 s, \qquad p <_1 q <_1 y, \quad \text{and} \quad \neg qrs$$

by the definition of $<_2$. By Lemma 6, xrs. Then A6 with y applied to xrs gives either yrs (hence $y <_1 r <_1 s$ by $r <_1 s$ and Lemma 6, and this with $p <_1 q <_1 y$ gives $q <_1 r <_1 s$, hence qrs, a contradiction) or xys or xry, and therefore $x \not\sim y$. □

Lemma 19 along with $I = sc(\prec)$ and the definition of $\sim$ in Section 4.1 gives

$$xIy \Leftrightarrow \text{not}(x \prec y \text{ or } y \prec x) \Leftrightarrow \text{not}(x \not\sim y) \Leftrightarrow x \sim y,$$

so that $I = \sim$. If $x \not\sim z$, so that not(xIz), then A6 with y applied to each of xzt, xtz, and txz implies either $x \not\sim y$ or $y \not\sim z$. Hence not(xIz) $\Rightarrow$ [not(xIy) or not(yIz)], and therefore $(xIy, yIz) \Rightarrow xIz$. Thus I is transitive, and the proof of Theorem 5 is complete.

The sufficiency proof of Theorem 6 can be done by showing that its conditions imply A2 and A6 (so that Theorem 5 applies) and also imply that, when T is nonempty, $x \neq y \Rightarrow (x \prec y$ or $y \prec x)$. Details are left to the reader.

5

Dimensionality and Other Parameters

Thus far we have focused on representations and characterizations of interval orders, interval graphs, and related binary relations. This chapter takes a closer look at particular facets of posets and interval orders, including height, width, dimensionality, breadth, depth, and magnitude. The bulk of the chapter is devoted to aspects of dimensionality and presents important theorems on dimensions of posets, semiorders, and interval orders. Observations on breadth and depth appear in the penultimate section, and the final section comments on numbers of interval orders and interval graphs.

5.1 PARAMETERS OF POSETS

Over the years many parameters have been defined for posets. We note six of these here. Although most research on these parameters has been done with finite posets, they will be defined for nonempty sets of arbitrary cardinality. Throughout this section, $(X, \prec)$ denotes a poset with X nonempty. It will be said to be finite if X is finite.

1. The *height* $H(X, \prec)$ of a finite poset $(X, \prec)$ is one less than the number of points in a maximum-cardinality chain in $(X, \prec)$. If there is a finite upper bound on the cardinalities of all chains in $(X, \prec)$, then $H(X, \prec)$ is one less than the cardinality of a maximum chain; otherwise, $H(X, \prec) = \sup\{|Y|: Y$ is a chain in $(X, \prec)\}$. Clearly, $H(X, \prec) = 0$ if and only if $\prec$ is empty, and $H(X, \succ) = n$ when $|X| = n + 1$ if and only if $(X, \prec)$ is a chain.

2. The *width* $W(X, \prec)$ of a poset is $\sup\{|A|: A$ is an antichain in $(X, \prec)\}$. Thus, $W(X, \prec) = |X|$ if and only if $\prec$ is empty—when X is finite, and $W(X, \prec) = 1$ if and only if $(X, \prec)$ is a chain. Chains and near-chains are tall and skinny. Posets with relatively few $\prec$ pairs are short and wide.

3. The *dimension* $D(X, \prec)$ of a poset is

$$\inf\{|\mathscr{L}|: \mathscr{L} \text{ is a family of linear orders on } X \text{ whose intersection equals } \prec\}.$$

Parameters of Posets 77

Since the cardinal numbers are well ordered, if $\{|\mathscr{L}|\}$ is nonempty, then it has a first (smallest) element. Hence, by Szpilrajn's extension theorem (Theorem 1.3) and the remark following its proof, $D(X, \prec)$ is well defined for all posets. Obviously, $D(X, \prec) = 1$ if and only if $W(X, \prec) = 1$, that is, $(X, \prec)$ is a linearly ordered set. Moreover, all weakly ordered sets that are not chains have dimension 2: for each equivalence class in $X/\sim$ (Theorem 1.2), use one linear extension of the class in one linear extension of $(X, \prec)$ and its dual in the other linear extension of $(X, \prec)$. In particular, $D(X, \prec) = 2$ if $\prec$ is empty and $|X| > 1$.

4. Let $x \preceq y$ mean that $x \prec y$ or $x = y$. A *join-independent* subset of a poset $(X, \prec)$ is a $Y \subseteq X$ such that for every $y \in Y$ there is a $g(y) \in X$ such that

$$\text{not}(y \preceq g(y)), \quad \text{and } Y\setminus\{y\} \preceq g(y) \text{ when } Y\setminus\{y\} \neq \emptyset.$$

Thus, Y is join-independent if for every $y \in Y$ there is a distinct $x = g(y)$ in X that is not above y but is above or equal to every other element in Y. The *breadth* $B(X, \prec)$ of a poset is $\sup\{|Y|: Y$ is a join-independent subset of $(X, \prec)\}$. $B(X, \prec) = 0$ if and only if $|X| = 1$, $B(X, \prec) = 1$ when $|X| > 1$ if and only if $\prec$ is a chain, and $B(X, \prec) = 2$ when $|X| > 1$ and $\prec$ is empty.

5. Given $(X, \prec)$ with $\sim = sc(\prec)$, define a binary relation $\subset_0$ on X by

$$x \subset_0 y \quad \text{if there are } a, b \in X \text{ such that } a \prec x \prec b \text{ and } y \sim \{a, x, b\}.$$

When $(X, \prec)$ is an interval order, $\subset_0$ is a partial order, but $\subset_0$ need be neither asymmetric nor transitive when $\prec$ is not an interval order. We have already encountered $\subset_0$ in the interval-order context of Section 2.2 (see Fig. 2.3) since it follows from the definitions that

$$x \subset_0 y \Leftrightarrow y \prec^- x \quad \text{and} \quad x \prec^+ y,$$

that is, that $F(x)$ begins after $F(y)$ begins and ends before $F(y)$ ends in every representation of $(X, \prec)$. For general posets, we define the *depth* of $(X, \prec)$ as $\sup\{|Y|: Y \subseteq X$ and Y is linearly ordered by $\subset_0\}$. Both chains and antichains have depth 1. All semiorders have depth 1.

6. The *magnitude* $m(X, \prec)$ of a poset is the cardinality of the set of maximal antichains in $(X, \prec)$. Magnitude was introduced in Section 2.3 by another definition but, as noted at the end of that section, the two definitions are commensurate when $(X, \prec)$ is a finite interval order. If $\prec = \emptyset$, then $m(X, \prec) = 1$; if $(X, \prec)$ is a chain, then $m(X, \prec) = |X|$.

Consistent with our usage in Section 3.2, we define the *magnitude* $m(X, \sim)$ of a reflexive graph $(X, \sim)$ as the cardinality of the set of maximal cliques in $(X, \sim)$. As in Corollary 3.1, $m(X, sc(\prec)) = m(X, \prec)$.

The posets shown in Fig. 5.1 provide examples for these six parameters. Details for posets I are: $H = 1$ since the maximum chain has two points;

$W = 3$ since there are (two) three-point antichains; $D = 3$ since the intersection of the three linear orders

$$xyazbc$$
$$zxbyca$$
$$yzcxab$$

equals $\{xa, sb, ya, yc, zb, zc\}$, and no two linear orders have the same intersection; $B = 3$ since $\{x, y, z\}$ is a three-point join-independent subset; depth $= 1$ since there is no three-point chain; $m = 5$ since the maximal antichains are abc, xyz, az, by, and cx. Details of the other cases are left to the reader.

A few more comments on Fig. 5.1 may be useful. Cases I, II, and IV are among the smallest posets with dimensionality exceeding 2. Indeed, no poset with fewer than six points has $D > 2$, and no semiorder with fewer than seven points has $D > 2$. Poset I is the only case in the figure whose breadth is 3 or more, and III is the only case with depth 2. In III, the middle point in the five-point chain stands in the relation C_0 to each of the points not in that chain.

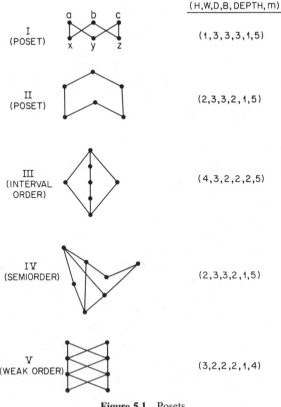

Figure 5.1 Posets.

5.2 DIMENSIONS OF POSETS

In this section we prove basic results for dimensions of posets and note several extensions proved elsewhere. We begin with the standard theorem of Dushnik and Miller (1941) which shows that every cardinal is the dimension of some poset.

Theorem 1 (Dushnik and Miller). *Let A be a set of cardinality $\alpha \geq 1$, and let $X = \{\{a\}: a \in A\} \cup \{A \setminus \{a\}: a \in A\}$. Then $D(X, \subset) = \alpha$.*

Proof. The theorem is easily verified for $\alpha \leq 2$. Assume henceforth that $\alpha \geq 3$, let $x, y, \ldots$, denote singleton subsets of A in X, and let $x^*, y^*, \ldots$ be their complements with respect to A. Then $X = \{x, y, \ldots\} \cup \{x^*, y^*, \ldots\}$, and $\subset$ consists of all pairs (x, y^*) with $x \neq y$. We prove $D(X, \subset) \geq \alpha$, then $D(X, \subset) \leq \alpha$.

Suppose $x \neq y$ and both $x^* < x$ and $y^* < y$ in a linear order $<$ on X. Then $x^* < y$ or $y^* < x$, so $<$ is not a linear extension of $\subset$. Since at most one (x^*, x) can be in a linear extension of $\subset$, and all such ordered pairs are needed to ensure that the intersection of a set of linear extensions equals $\subset$, $D(X, \subset) \geq \alpha$.

For every x, let $<_x$ be a linear extension of $\subset$ that has $x^* <_x x$ along with $y <_x x^*$ and $x <_x y^*$ for all $y \neq x$. The existence of $<_x$ as a linear extension of $\subset$ is transparent. Clearly, $|\{<_x: x \in X\}| = \alpha$. We claim that $(\cap <_x) = \subset$, so that $D(X, \subset) \leq \alpha$. Indeed, for x^* and y^* when $x \neq y$, $x^* <_x x <_x y^*$ and $y^* <_y y <_y x^*$; similarly, $y <_x x^* <_x x$ and $x <_y y^* <_y y$ for x versus y; and finally, for x versus x^*, $x^* <_x x$ and $x <_y y^* <_y x^*$ whenever $y \neq x$. □

For finite $n \geq 3$, Theorem 1 shows that there are posets with $2n$ points and dimension n. In particular, the family of 1-element and $(n-1)$-element subsets of an n-element set ordered by proper inclusion has dimension n. Case I of Fig. 5.1 shows this poset for $n = 3$. Bogart and Trotter (1973) prove that every six-point poset with dimension 3 is isomorphic to I or II or the dual of II in Fig. 5.1. They also prove that if $n \geq 4$ and if a $2n$-point poset has dimension n, then it is isomorphic to the family of 1-element and $(n-1)$-element subsets of an n-set ordered by $\subset$. Hiraguchi's theorem, stated later as Theorem 7, shows that no poset with fewer than $2n$ points can have dimension n when $n \geq 3$. Thus, the results of Bogart and Trotter identify all minimum-cardinality posets of finite dimension $n \geq 3$.

Komm (1948) shows that $D(X, \subset)$ in Theorem 1 cannot be increased if other subsets of A are added to X. Let 2^A be the set of all subsets of A.

Theorem 2 (Komm). *Let A be a set of cardinality $\alpha \geq 1$. Then $D(2^A, \subset) = \alpha$.*

Proof. In view of Theorem 1 and its proof, we need only show that $D(2^A, \subset) \leq \alpha$ when $\alpha = |A| \geq 3$. Construct α linear extensions of $\subset$ on 2^A as follows. For each $a \in A$, partition $A \setminus \{a\}$ into $A_1(a) = \{B \subseteq A: B \neq \{a\}$ and $\{a\} \not\subset B\}$ (note that $\varnothing \in A_1(a)$) and $A_2(a) = \{B \subseteq A: \{a\} \subset B\}$, let $P_i(a)$ for $i \in \{1, 2\}$ be the restriction of $(2^A, \subset)$ to $A_i(a)$, let $L_i(a)$ for $i \in \{1, 2\}$ be a linear extension of $P_i(a)$, and let $<_a$ be a linear extension of $(2^A, \subset)$ for which $L_1(a) <_a \{a\} <_a L_2(a)$, where $L_i(a)$ denotes the linear order on $A_i(a)$ just defined. It is routine to verify that the intersection of the $<_a$ equals $\subset$ on 2^A. □

The preceding theorems identify posets of specific dimensionality. The rest of this section considers bounds on $D(X, \prec)$. We shall establish several upper bounds, the best known of which is Hiraguchi's theorem: $D(X, \prec) \leq \frac{1}{2}|X|$ when $|X| \geq 4$. Other upper bounds due to Hiraguchi (1955) and Dilworth (1950) are presented in Theorems 3 and 5, and two important upper bounds from Trotter (1975) appear in Theorems 4 and 6. Trotter also shows for finite posets that $D(X, \prec) \leq W(X \setminus E, \prec) + 1$ when E is a set of maximal (minimal) points in $(X, \prec)$.

Apart from the obvious lower bound on $D(X, \prec)$ in Theorem 3(a), the only direct lower-bound result that we shall present comes in the final theorem of the section, Theorem 8, due to Baker (1961). It says that the dimension of a poset is at least as great as its breadth. Since $B(X, \subset) = \alpha$ in the context of Theorems 1 and 2, Baker's lower bound is tight in some cases.

For convenience in what follows, we take

$$D(\varnothing, \prec) = 0,$$
$$\sim \ = sc(\prec),$$

and will say that $\mathscr{L}$ realizes $(X, \prec)$ when $\mathscr{L}$ is a set of linear extensions of $\prec$ whose intersection equals $\prec$. Thus $D(X, \prec) = \min\{|\mathscr{L}|: \mathscr{L}$ realizes $(X, \prec)\}$. In addition, linear orders will sometimes be written by juxtaposition of subsets and/or single elements in X. For example, if A_1, A_2, and A_3 are disjoint linearly ordered subsets of X and $X \setminus (\cup A_i) = \{x\}$, then $A_1 x A_3 A_2$ denotes the linear order $\prec^*$ on X in which $\prec^*$ coincides with the order on each A_i, and $A_1 \prec^* x \prec^* A_3 \prec^* A_2$.

We shall also have occasion to use close relatives of Spzilrajn's extension theorem from Bogart (1973b), as follows.

Lemma 1. *If $(X, \prec)$ is a poset, $\varnothing \subset Y \subset X$, and $(Y, \prec_1)$ is a linear extension of $(Y, \prec)$, then there is a linear extension $(X, \prec^*)$ of $(X, \prec)$ with $\prec_1 \subset \prec^*$.*

Lemma 2. *If $(X, \prec)$ is a poset and $(C, \prec)$ is a chain in $(X, \prec)$, then $t(\prec \cup \{(x, c): x \in X \setminus C, c \in C, x \sim c\})$ and $t(\prec \cup \{(c, x): x \in X \setminus C, c \in C, x \sim c\})$ are partial orders on X.*

Remark 1. Within the context of Lemma 2, we shall say that $\prec^*$ is a *lower extension* of $\prec$ for C when $(X, \prec^*)$ is a linear extension of $(X, \prec)$ that has

Dimensions of Posets

$x \prec^* c$ whenever $x \in X \setminus C$, $c \in C$, and $x \sim c$. An *upper extension* of $\prec$ for C is defined dually, with $c \prec^* x$ in place of $x \prec^* c$. Lemma 2 says that there are both lower and upper extensions of $\prec$ for C when account is taken of Theorem 1.3.

Remark 2. Bogart (1973b) also notes that if C_1 and C_2 are disjoint chains in $(X, \prec)$ and $C_1 \sim C_2$, then there is a linear extension $(X, \prec^*)$ of $(X, \prec)$ that has $x \prec^* c_1$ whenever $x \in X \setminus C_1$, $c_1 \in C_1$, and $x \sim c_1$, and $c_2 \prec^* x$ whenever $x \in X \setminus C_2$, $c_2 \in C_2$, and $x \sim c_2$. (Therefore $C_2 \prec^* C_1$.) This mixed lower-C_1 and upper-C_2 extension is in effect a corollary of Lemma 2.

Proofs. Given the hypotheses of Lemma 1, it is easily verified that

$$\prec \cup \{(a,b): a \leqq x \prec_1 y \leqq b \text{ for some } x, y \in Y\}$$

is irreflexive and transitive, hence a partial order, so by Szpilrajn's theorem there is a linear extension of $\prec$ that includes $\prec_1$.

Given the hypotheses of Lemma 2, it is routine to show that

$$t(\prec \cup \{(x,c): x \in X \setminus C, c \in C, x \sim c\})$$
$$= \prec \cup \{(x,y): x \in X \setminus C, x \sim c \leqq y \text{ for some } c \in C\},$$

and then that the latter expression is a partial order on X. The other part of Lemma 2 is proved similarly. □

The first of our bounding theorems for D is

Theorem 3 (Hiraguchi). *Suppose $(X, \prec)$ is a poset, $x \in X$, C_i is a chain of $(X, \prec)$, and $(X_i, \prec_i)$ is a poset with $X_i \subseteq X$ for $i = 1, 2$. Then:*
  (a) $D(X \setminus \{x\}, \prec) \leq D(X, \prec) \leq D(X \setminus \{x\}, \prec) + 1$.
  (b) $D(X, \prec) = \max\{D(X_1, \prec_1), D(X_2, \prec_2), 2\}$ *if* $\prec = \prec_1 \cup \prec_2$, $X_1 \cap X_2 = \emptyset$, *and neither* $\prec_i$ *is empty.*
  (c) $D(X, \prec) \leq D(X \setminus (C_1 \cup C_1), \prec) + 2$ *when* $C_1 \cap C_2 = \emptyset$ *and* $C_1 \sim C_2$.
  (d) $D(X, \prec) \leq D(X \setminus C_1, \prec) + 2$.

Proof. (a) Since deletion of x from each chain in a set that realizes $(X, \prec)$ leaves a set that realizes $(X \setminus \{x\}, \prec)$, $D(X \setminus \{x\}, \prec) \leq D(X, \prec)$. For the other part of (a), let $\mathscr{L}$ realize $(X \setminus \{x\}, \prec)$, fix $L \in \mathscr{L}$, and let $A = \{y \in X: x \prec y\}$ and $B = \{y \in X: y \prec x\}$. Form $|\mathscr{L}| + 1$ chains on x that realize $(X, \prec)$ as follows. For each chain L' in $\mathscr{L}$ other than L, place x so that $BL'x$ when $B \neq \emptyset$ and $xL'A$ when $A \neq \emptyset$. This is always possible since if A and B are not empty, then $B \prec A$ and hence $BL'A$. We form two chains on X, L_1 and L_2, from L on $X \setminus \{x\}$ as follows. Let

$$C(A) = \{y \neq x: y \sim x \text{ and } aLy \text{ for some } a \in A\} \text{ if } A \neq \emptyset,$$

$$C(B) = \{y \neq x: y \sim x \text{ and } yLb \text{ for some } b \in B\} \text{ if } B \neq \emptyset.$$

If $A = \emptyset$, form L_1 by putting x at the top of L; otherwise, move all points in $C(A)$ (if any) in their order in L immediately below A, then insert x immediately below A (and above $C(A)$ thus moved) to yield L_1. If $B = \emptyset$, form L_2 by putting x at the bottom of L; otherwise, move all points in $C(B)$ (if any) in their order in L immediately above B, then insert x immediately above B to yield L_2. It is easily seen that $|\mathcal{L}| + 1$ chains on X thus formed realize $(X, \prec)$.

(b) Let $(X_i, \prec_i)$ for $i = 1, 2$ be posets within $(X, \prec)$ such that $X_1 \cap X_2 = \emptyset$ and $\prec_1 \cup \prec_2 = \prec$ (if possible), and let $X_3 = X \setminus (X_1 \cup X_2)$. If each $(X_i, \prec_i)$ is a chain, it is easily seen that $D(X, \prec) = 2$. Assume henceforth that at least one of the $(X_i, \prec_i)$ is not a chain, so that $\min\{D(X_1, \prec_1), D(X_2, \prec_2)\} \geq 2$, and suppose for definiteness that

$$\alpha_1 = D(X_1, \prec_1) \geq D(X_2, \prec_2) = \alpha_2.$$

Let $\mathcal{L}_i$ with $|\mathcal{L}_i| = \alpha_i$ realize $(X_i, \prec_i)$, and if X_3 is not empty let L_3 be a linear order of X_3. Since $\alpha_1 \geq \alpha_2$, there is a mapping g from $\mathcal{L}_1$ onto $\mathcal{L}_2$. Fix L in $\mathcal{L}_1$. Form a linear order on X by placing L_3 above L above $g(L)$. For every other L' in $\mathcal{L}_1$, form a linear order on X by placing $g(L')$ above L' above $d(L_3)$. This gives α_1 chains on X that realize $(X, \prec)$. Since α_1 chains are needed to realize $(X, \prec_1)$, it follows that $D(X, \prec) = \alpha_1$.

(c) Assume that C_1 and C_2 are chains in $(X, \prec)$ with $C_1 \cap C_2 = \emptyset$ and $C_1 \sim C_2$. If $X = C_1 \cup C_2$, the conclusion of (c) follows from (b). Suppose henceforth that $C_1 \cup C_2 \subset X$, let $\alpha = D(X \setminus (C_1 \cup C_2), \prec)$, and let $\mathcal{L}$ realize $(X \setminus (C_1 \cup C_2), \prec)$ with $|\mathcal{L}| = \alpha$. We form $\alpha + 2$ chains on X that realize $(X, \prec)$ as follows. For each $L \in \mathcal{L}$ let L^* be a linear extension on X of $\prec \cup L$ as guaranteed by Lemma 1. Using Lemma 2 (see Remark 2), let L_1 be a linear extension of $\prec$ on X that has xL_1c_1 whenever $x \in X \setminus C_1$, $c_1 \in C_1$, and $x \sim c_1$, and has c_2L_1x whenever $x \in X \setminus C_2$, $c_2 \in C_2$, and $x \sim c_2$. Also let L_2 be a linear extension of $\prec$ on X that has xL_2c_2 whenever $x \in X \setminus C_2$, $c_2 \in C_2$, and $x \sim c_2$, and has c_1L_2x whenever $x \in X \setminus C_1$, $c_1 \in C_1$, and $x \sim c_1$. For L_1 and L_2, $C_2L_1C_1$ and $C_1L_2C_2$. It is easily seen that the $\alpha + 2$ chains thus formed on X realize $(X, \prec)$.

(d) This proof is similar to the proof of (c) with $C_2 = \emptyset$. In the latter part of the preceding paragraph, L_1 is a lower extension for C_1, and L_2 is an upper extension of C_1. (See Remark 1.) $\square$

Our next upper bound is based on what is left after an antichain is removed from $(X, \prec)$.

Theorem 4 (Trotter). *Suppose A is an antichain in a poset $(X, \prec)$ with $|X \setminus A| \geq 2$. Then*

$$D(X, \prec) \leq |X \setminus A|.$$

Proof. Given the hypotheses of the theorem, suppose first that $X \setminus A = \{x, y\}$. If $x \prec y$ and no $a \in A$ has $a \prec x$ or $y \prec a$, let $A_1 = \{a: x \prec a \prec y\}$, $A_2 = \{a \in A: x \sim a \prec y\}$, $A_3 = \{a \in A: x \prec a \sim y\}$, and $A_4 = \{a: x \sim a \sim y\}$. Let L_i be a linear order of A_i when $A_i \neq \emptyset$. Then $(X, \prec)$ is realized by the chains $L_4 x L_3 L_1 L_2 y$ and $d(L_2) x d(L_1) y d(L_3) d(L_4)$. If $a \prec x \prec y$ for some $a \in A$, let $A_1 = \{a: a \prec x\}$, $A_2 = \{a \in A: x \sim a \prec y\}$, and $A_3 = \{a: x \sim a \sim y\}$. With L_i a linear order of A_i, the chains $L_1 x L_2 y L_3$ and $d(L_3) d(L_2) d(L_1) xy$ realize $(X, \prec)$ in this case. A similar result holds if $x \prec y \prec a$ for some $a \in A$. Related computations for $x \sim y$ show in general that

$$D(X, \prec) \leq 2 \quad \text{if } |X \setminus A| = 2.$$

Suppose next that $|X \setminus A| > 2$. Let Y be a subset of $X \setminus A$ that contains all but two points in $X \setminus A$. Then $D(X \setminus Y, \prec) \leq 2$ by the result just proved, and $D(X, \prec) \leq D(X \setminus Y, \prec) + |Y|$ by Theorem 2(a), so

$$D(X, \prec) \leq 2 + |Y| = |X \setminus A|. \qquad \square$$

Remark 3. The final sentence of the preceding proof glosses over the fact that

$$D(X, \prec) \leq D(X \setminus Y, \prec) + |Y|.$$

This clearly follows from $D(X, \prec) \leq D(X \setminus \{x\}, \prec) + 1$ in Theorem 3(a) if Y is finite. If Y is infinite, suppose $\mathscr{L}$ with $|\mathscr{L}| = \alpha = D(X \setminus Y, \prec)$ realizes $(X \setminus Y, \prec)$. By Lemma 1, for each $L \in \mathscr{L}$ there is a linear extension of $\prec$ on X that includes L. In addition, for each $y \in Y$ there is a pair of linear extensions of $\prec$ on X such that, in one extension, $y \prec^* x$ for every $x \neq y$ for which $x \sim y$, and, in the other extension, $x \prec^* y$ for every $x \neq y$ for which $x \sim y$. (Proof left to the reader.) Consequently, $\prec$ equals the intersection of $\alpha + 2|Y|$ linear extensions of $(X, \prec)$. But, since $|Y|$ is infinite, $2|Y| = |Y|$, so $D(X, \prec) \leq D(X \setminus Y, \prec) + |Y|$.

The next two upper bounds on D use the width function W in somewhat different ways. Since Dilworth (1950) mentions the first result in his initial footnote, I have attributed it to him. The second result is discussed extensively in Trotter (1974, 1975).

Theorem 5 (Dilworth). *If the width of poset $(X, \prec)$ is finite, then $D(X, \prec) \leq W(X, \prec)$.*

Theorem 6 (Trotter). *If $A \subset X$ is an antichain in a poset $(X, \prec)$ and the width of $(X \setminus A, \prec)$ is finite, then*

$$D(X, \prec) \leq 2W(X \setminus A, \prec) + 1.$$

Remark 4. The ensuing proofs of Theorems 5 and 6 use Dilworth's decomposition theorem, Theorem 1.6. Since Perles (1963b) shows that the infinite-width analog of Dilworth's theorem is not true, the proof method used below does not apply to infinite W. I do not know whether Theorems 5 and 6 remain true when the finiteness restriction on W is removed. Perles's example, which shows that there are posets of arbitrary cardinality α which have no infinite antichains but cannot be decomposed in the manner of Theorem 1.6 into fewer than α chains, does not disprove the generalization of Theorem 5 since its posets have dimensionality 2.

Proofs. For Theorem 5, let A be a maximum antichain of $(X, \prec)$ with $|A| = W(X, \prec) = n$ finite. By Theorem 1.6, $X = \cup X_i$ for chains $(X_1, \prec_1), \ldots, (X_n, \prec_n)$ in $(X, \prec)$. For each i let $\prec_i^*$ be an upper extension (Remark 1) of $\prec$ for X_i. Then, since $x \sim y$ and $x \neq y$ give $x \prec_i^* y$ for $x \in X_i$ and $y \prec_j^* x$ for $y \in X_j$, it follows that $\prec = (\cap \prec_i^*)$, so $D(X, \prec) \leq n$.

For Theorem 6, suppose A is an antichain in $(X, \prec)$, $A \neq X$, and $W(X \setminus A, \prec) = n$ is finite. By Theorem 1.6, $X \setminus A = \cup X_i$ for n chains $(X_i, \prec_i)$ in $(X \setminus A, \prec)$. For each i, let L_i and U_i be lower and upper extensions of $\prec_i$ for X_i (to all of X). Also let H_0 on A be the dual of L_n restricted to A, and let H be a linear extension of $\prec$ whose restriction to A is H_0. Since it is easily seen that $H \cap (\cap L_i) \cap (\cap U_i) = \prec$, we have $D(X, \prec) \leq 2W(X \setminus A, \prec) + 1$. □

We now come to Hiraguchi's theorem. His own proof of the theorem was very involved and others, including Bogart (1973b), developed alternate proofs. The simple proof used here is due to Trotter (1975) following a suggestion of Bogart.

Theorem 7 (Hiraguchi). $D(X, \prec) \leq \frac{1}{2}|X|$ *for every poset with $|X| \geq 4$.*

Proof. The result is trivially true if X is infinite, since then $D(X, \prec) \leq |X \times X| = |X| = \frac{1}{2}|X|$. For finite X with cardinality 4 or more, the desired inequality follows immediately from Theorems 4 and 5. □

Our final theorem of the section is Baker's (1961) lower-bound result.

Theorem 8 (Baker). $B(X, \prec) \leq D(X, \prec)$ *for every poset.*

Proof. The proof is basically the same as the $D \geq \alpha$ proof of Theorem 1. Theorem 8 is obviously true if $B \leq 2$. If $B(X, \prec) > 2$, and Y is any join-independent subset of $(X, \prec)$ with three or more points, then for each $y \in Y$ there is a $y^* \neq y$ in X such that $Y \setminus \{y\} \prec y^* \sim y$. It follows that at most one y^*

precedes y for $y \in Y$ in a linear extension of $(X, \prec)$. Hence $|Y| \leq D(X, \prec)$ for every join-independent subset Y of $(X, \prec)$, and therefore $B(X, \prec) \leq D(X, \prec)$. □

For finite X, $B(X, \prec) \leq D(X, \prec) \leq W(X, \prec)$ by Theorems 5 and 8, with $B = W$ in the setting of Theorem 1.

5.3 POSETS OF DIMENSION 2

The characterization of posets with dimension 1 is elementary since they are precisely the linearly ordered sets. In this section we consider the posets of dimension 2. Our first result presents alternative characterizations of $D = 2$ due mainly to Dushnik and Miller (1941), supplemented by a few observations from Baker, Fishburn, and Roberts (1972). We then show that $D = 2$ for two special classes of interval orders. Finally, it is noted that $D = 2$ cannot be characterized by a finite list of forbidden posets.

Throughout the section we take $\sim \, = sc(\prec)$, and define $\sim°$ on X by

$$x \sim° y \quad \text{if } x \sim y \text{ and } x \neq y.$$

The relation $\sim°$ is often referred to as the incomparability relation of $(X, \prec)$, and $(X, \sim°)$ is its incomparability graph. If $(X, \prec^*)$ is a linear extension of a poset $(X, \prec)$, we shall say that it is *nonseparating* if, for all $x, y, z \in X$,

$$(x \prec^* y \prec^* z, x \sim° y, y \sim° z) \Rightarrow x \sim° z.$$

Theorem 9 (Dushnik and Miller). *Suppose $(X, \prec)$ is a poset that is not a chain. Then the following are mutually equivalent*:

(a) $D(X, \prec) = 2$.

(b) $(X, \prec)$ has a nonseparating linear extension.

(c) There is a partial order P on X such that $P \cup d(P) = \, \sim°$; that is, $(X, \sim°)$ is a comparability graph.

(d) There exists a 1-1 mapping F from X into closed intervals of a linearly ordered set $(Y, <_0)$ such that, for all $x, y \in X$,

$$x \prec y \Leftrightarrow F(x) \subset F(y).$$

(e) There exists a 1-1 mapping F from X into closed intervals of a linearly ordered set $(Y, <_0)$ such that, for all $x, y \in X$,

$$x \prec y \Leftrightarrow \left[\inf F(x) <_0 \inf F(y) \text{ and } \sup F(x) <_0 \sup F(y)\right].$$

Proof. Assume that $(X, \prec)$ is a poset that is not a chain. We prove that (a) $\Rightarrow$ [(d) & (e)], [(d) or (e)] $\Rightarrow$ (c), (c) $\Rightarrow$ (b), and (b) $\Rightarrow$ (a).

(a) $\Rightarrow$ [(d) and (e)]. Given (a), let $\prec = L_1 \cap L_2$, where $\{L_1, L_2\}$ realizes $\prec$. Let X' be a disjoint copy of X, with $x' \in X'$ corresponding to $x \in X$. For (d), let L'_2 on X' be the copy of $d(L_2)$ on X, and let $(Y, <_0) = (X \cup X', L'_2 L_1)$ with $F(x) = [x', x]$. Then F is 1-1 and the conclusion of (d) follows. For example, $x \prec y \Leftrightarrow (xL_1 y, xL_2 y) \Leftrightarrow (xL_1 y, yd(L_2)x) \Leftrightarrow (y'L'_2 x', xL_1 y) \Leftrightarrow y' <_0 x' <_0 x <_0 y \Rightarrow [x', x] \subset [y', y]$. For (e), do exactly the same thing except take L'_2 on X' as the copy of L_2 on X. The conclusion of (e) then follows for this version of $(Y, <_0)$.

[(d) or (e)] $\Rightarrow$ (c). If (d) holds then P defined by

$$xPy \Leftrightarrow (x \neq y, \inf F(x) \leq_0 \inf F(y), \sup F(x) \leq_0 \sup F(y))$$

is easily seen to be a partial order such that $P \cup d(P) = \sim^\circ$, If (e) holds, then P defined by strict inclusion on the intervals $F(x)$ for (e) is a partial order for which $P \cup d(P) = \sim^\circ$.

(c) $\Rightarrow$ (b). Assume that (X, P) is a poset with $P \cup d(P) = \sim^\circ$. Let $\prec^* = \prec \cup P$. It is straightforward to show that $\prec^*$ is a linear order on X. Moreover, it is nonseparating, for $(x \prec^* y \prec^* z, x \sim^\circ y, y \sim^\circ z) \Rightarrow (xPy, yPz) \Rightarrow xPz \Rightarrow x \sim^\circ z$.

(b) $\Rightarrow$ (a). Assume that $\prec^*$ is a nonseparating linear extension of $\prec$, and let $P = \prec^* \setminus \prec$. If xPy and yPz, then $x \sim^\circ y$ and $y \sim^\circ z$, and, since $x \prec^* z$, nonseparation implies $x \sim^\circ z$, so xPz. Therefore P is a partial order. Since $\prec \cup d(P)$ is easily seen to be a linear extension of $\prec$ along with $\prec \cup P = \prec^*$, $D(X, \prec) \leq 2$ since $\prec = [\prec \cup P] \cap [\prec \cup d(p)]$. Because $(X, \prec)$ is not a chain, $D(X, \prec) = 2$. □

Our two theorems for interval orders of dimension 2 are due, respectively, to Rabinovitch (1978a) and Baker, Fishburn, and Roberts (1970). The first says that interval orders of height 1 have $D \leq 2$, in sharp contrast to the fact (Theorem 1) that there are posets of height 1 with arbitrarily large dimensionality. The second shows that a special class of interval orders that is more general than the class of weak orders has no member whose dimension exceeds 2. We refer to an interval order $(X, \prec)$ as *strong* if, for all $a, b, x, y \in X$,

$$(a \prec x, b \prec y, a \sim b, x \sim y) \Rightarrow (a \prec y \text{ and } b \prec x).$$

This condition strengthens (2.1) of Section 2.1, which requires only $a \prec y$ or $b \prec x$ on the right hand side.

Theorem 10 (Rabinovitch). *Suppose $(X, \prec)$ is an interval order with $H(X, \prec) = 1$ and $|X| \geq 3$. Then $D(X, \prec) = 2$.*

Theorem 11 (Baker, Fishburn, and Roberts). *Suppose $(X, \prec)$ is a strong interval order that is not a chain. Then $D(X, \prec) = 2$.*

Proofs. The hypotheses of Theorem 10 imply that $D(X, \prec) \geq 2$. To show that $D(X, \prec) = 2$ in this case, let $A = \{x: y \prec x \text{ for some } y\}$, $B = \{x: x \prec y \text{ for some } y\}$, and $U = \{x: x \sim X\}$. Because $H(X, \prec) = 1$, A, B, and U partition X. We shall ignore U in what follows since the linear extensions of $\prec$ on $A \cup B$ obtained below can clearly be augmented by linear orders on U to give $D = 2$.

Let $\prec_1$ be a linear extension of $\prec$ on $A \cup B$ that has $a \prec_1 b$ whenever $a \in A$, $b \in B$, and $a \sim b$. The existence of such an extension can be seen by using the restrictions on A and B of the order $\prec_0^*$ suggested by Theorem 2.4, plus $\prec$ on $A \cup B$ and $\{(a, b): a \in A, b \in B, a \sim b\}$. A similar proof of such a $\prec_1$ appears later as a special case of Theorem 13. In addition, let $\prec_2 = d(B, \prec_1)d(A, \prec_1)$. It then follows that $\prec = \prec_1 \cap \prec_2$, and the proof of Theorem 10 is complete.

For Theorem 11, suppose $(X, \prec)$ is a strong interval order that is not a chain, so $D(x, \prec) \geq 2$. Given the weak orders $\prec^- = (\sim)(\prec)$ and $\prec^+ = (\prec)(\sim)$ of Theorem 2.2, $x \prec y \Rightarrow (x \prec^- y, x \prec^+ y)$ and, by the strong interval order property, $(x \prec^- y, x \prec^+ y) \Rightarrow x \prec y$:

$$(x \prec^- y, x \prec^+ y) \Rightarrow (x \sim a \prec y, x \prec b \sim y) \Rightarrow (a \prec b \text{ and } x \prec y).$$

Therefore $\prec = (\prec^- \cap \prec^+)$. When $\prec^-$ is augmented by all (x, y) such that $x \sim^- y$ and $y \prec^+ x$, and $\prec^+$ is augmented by all (x, y) such that $x \sim^+ y$ and $y \prec^- x$, the intersection of the augmented orders equals $\prec^- \cap \prec^+$. Moreover, the augmented orders are linear (see Theorem 2.2) unless there are distinct x and y with $x \approx y$, and if $x \approx y$ then x and y are in the same equivalence class in augmented $\prec^-$ and in the same equivalence class in augmented $\prec^+$. Taking dual orders within these classes, we get two linear orders whose intersection equals $\prec$. □

Our final observation for $D = 2$, also from Baker, Fishburn, and Roberts (1970, 1972) and, independently, from Harzheim (1970), says that $D = 2$ for nonchain posets cannot be determined by checking that no restriction of $(X, \prec)$ is isomorphic to some poset in a fixed finite set of posets that have $D > 2$. It is easily seen that a similar result holds for $D = n$ (or $D \leq n$), but we omit the details, which are noted in Baker, Fishburn, and Roberts (1970, pp. 20–21) and Baker (1961).

Theorem 12. *There is no finite set $\mathcal{P}$ of posets such that, for every poset $(X, \prec)$ that is not a chain, $D(X, \prec) = 2$ if and only if no restriction of $(X, \prec)$ is isomorphic to a poset in $\mathcal{P}$.*

Proof. Suppose $\mathcal{P}$ is a set of posets such that every nonchain poset has $D = 2$ if and only if it has no restriction isomorphic to a poset in $\mathcal{P}$. To show that $\mathcal{P}$ must be infinite, consider the fence $(\prec^m)$ and crown $(\prec_m)$ posets of height 1 for $m \geq 3$ in Fig. 5.2. Since $\prec^m$ is realized by $\{b_1a_1b_2a_2 \ldots b_ma_m,$

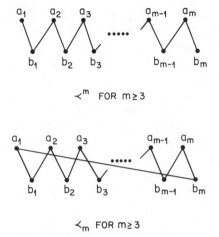

Figure 5.2 Fences (top) and crowns (bottom).

$b_m b_{m-1} a_m b_{m-2} a_{m-1} b_{m-3} \ldots b_2 a_3 b_1 a_2 a_1\}$, it has $D = 2$. However, $D = 3$ for every crown $\prec_m$. Crown $\prec_3$ is isomorphic to poset I in Fig. 5.1 ($D = 3$ by Theorem 1) and it is easily checked for general $\prec_m$ that $D = 3$. Since $\prec_m$ and $\prec^m$ have precisely the same posets as proper restrictions, and since these all have $D \leq 2$ because $D = 2$ for $\prec^m$, it follows that every $\prec_m$ must be in $\mathscr{P}$. □

Trotter (1974) defines $(X, \prec)$ as *irreducible* if $D(X \setminus \{x\}, \prec) < D(X, \prec)$ for every $x \in X$. The crowns in Fig. 5.2 are irreducible posets of height 1. Trotter proves that there exist irreducible posets of arbitrarily large finite height. His examples show that the inequality $D(X, \prec) \leq 2W(X \setminus A, \prec) + 1$ of Theorem 6 is actually an equality for the cases considered. Additional results on irreducible posets of dimension 3 are presented by Kelly and Rival (1975).

5.4 DIMENSIONS OF SEMIORDERS

Since interval orders of height 1 are semiorders, Theorem 10 says that $D \leq 2$ for every semiorder of height 1. We also know by case IV of Fig. 5.1 that some semiorders of height 2 have dimension 3. Moreover, as we shall see in the next section, there are interval orders with arbitrarily large finite dimensions.

In view of these observations, the surprising fact is that $D(X, \prec) \leq 3$ for *all* semiorders. This was proved by Rabinovitch (1978a) for finite X, and will be proved here for X of arbitrary cardinality.

Rabinovitch's semiorder theorem will be considered after we note a necessary preliminary result from Rabinovitch (1978b) that gives an important characterization of interval orders. In proving the latter theorem, we shall use lexicographic extensions of the weak orders $\prec^- = (\sim)(\prec)$ and $\prec^+ = (\prec)(\sim)$

of Theorem 2.2 defined as follows:

$$x <^- y \text{ if } [x \prec^- y \text{ or } (x \sim^- y \text{ and } x \prec^+ y)],$$
$$x <^+ y \text{ if } [x \prec^+ y \text{ or } (x \sim^+ y \text{ and } x \prec^- y)].$$

Rabinovitch (1978a) proves that an interval order is a semiorder if and only if $<^- = <^+$ (see Theorems 2.2 and 2.4), in which case $<^- = <^+ = \prec^- \cup \prec^+$. For general interval orders, if no two points in X are equivalent in the sense of $\approx$ (Section 2.2), then $(X, <^-)$ and $(X, <^+)$ are linear extensions of $(X, \prec)$ since they completely order elements within the classes of $X/\sim^-$ and $X/\sim^+$. (The lexicographic extensions just defined are *not* the same as the augmented orders used in the proof of Theorem 11. For example, the augmented order for $\prec^-$ used there has x less than y if $x \prec^- y$ or $(x \sim^- y$ and $y \prec^+ x)$.)

For any disjoint subsets A and B of a poset $(X, \prec)$ we shall let $[A|B]$ denote a (not necessarily specific) linear extension $(A \cup B, \prec^*)$ of $(A \cup B, \prec)$ for which

$$a \prec^* b \text{ whenever } a \in A, b \in B \text{ and } a \sim b,$$

where as usual $\sim = sc(\prec)$. If either A or B is empty, such an extension is trivially assured by Szpilrajn's extension theorem. If $A = B = \emptyset$, we shall say that $[A|B]$ exists. For nonempty $A \cup B$, we shall say that $[A|B]$ exists if there is a linear extension of $(A \cup B, \prec)$ as just noted.

Theorem 13 (Rabinovitch). *A poset $(X, \prec)$ is an interval order if and only if $[A|B]$ exists for every two disjoint subsets A and B of X.*

Proof. If poset $(X, \prec)$ is not an interval order, then there are $a, b, x, y \in X$ such that $a \prec x$, $b \prec y$, $a \sim y$, and $b \sim x$. Let $A = \{x, y\}$ and $B = \{a, b\}$. Then $[A|B]$ requires $x \prec^* b$ and $y \prec^* a$, which are incompatible with $a \prec^* x$ and $b \prec^* y$ for a linear extension. Hence $[A|B]$ does not exist.

To prove the converse, assume without loss of generality that $x \neq y \Rightarrow x \not\approx y$ and that A and B are nonempty disjoint subsets of interval order $(X, \prec)$. We are to show that $[A|B]$ exists. Given the linear extensions $<^-$ and $<^+$ of $\prec$ as defined above, let

$$(\prec^* \text{ on } A \cup B) = (\prec \text{ on } A \cup B) \cup (<^- \text{ on } A) \cup (<^+ \text{ on } B)$$
$$\cup \{(a, b): a \in A, b \in B, a \sim b\}.$$

We claim that $\prec^*$ linearly orders $A \cup B$, hence that $[A|B]$ exists. Since $\prec^*$ is clearly irreflexive and complete, only transitivity needs to be checked. This is routine. For example, if $a \sim b$, and $b \prec a'$, then $a <^- a'$, so $a \prec^* a'$. Or, if $a' <^- a$ and $a \sim b$, then not$(b \prec a')$ since otherwise $a <^- a'$; so either $a' \prec b$ or $a' \sim b$, and therefore $a' \prec^* b$. Other cases are similar. □

Theorem 14 (Rabinovitch). *Suppose $(X, \prec)$ is a semiorder. Then $D(X, \prec) \leq 3$.*

Proof. Let $(X, \prec)$ be a semiorder with $H(X, \prec) > 1$, since otherwise the desired conclusion is given by Theorem 10. Because dual linear orders in equivalence classes ($\approx$) can be used in different linear extensions of $X/\approx$, we assume without loss of generality that $x \neq y \Rightarrow x \not\approx y$. Let F be a representation of $(X, \prec)$ of the type assured by Theorem 2.7 with $x \prec y \Leftrightarrow F(x) <_0 F(y)$, and $\inf F(x) <_0 \inf F(y) \Leftrightarrow \sup F(x) <_0 \sup F(y)$. Because $x \neq y \Rightarrow x \not\approx y$, $<_0$ on $\{\inf F(x)\}$ is a linear order and is the same as $<_0$ on $\{\sup F(x)\}$. (See also Theorem 3.10, and note that $<_0$ on $\{\inf F(x)\}$ is exactly the same as $<^-$ on X or $<^+$ on X under the obvious isomorphism.)

Let X' be a component of $(X, \sim)$. The union of the $F(x)$ for $x \in X'$ is itself an interval of the linearly ordered set used to define F. Moreover, the total intervals for two such components of $(X, \sim)$ are disjoint. Hence it will suffice to prove that $D(X', \prec) \leq 3$ for an arbitrary component X' of $(X, \sim)$.

For notational convenience, assume henceforth that $(X, \sim)$ is connected. Let $(Y, \prec)$ be a maximal chain in $(X, \prec)$ as assured by Kuratowski's lemma in Section 1.4. Suppose Y can be partitioned into nonempty Y_1 and Y_2 with $Y_1 \prec Y_2$ such that either Y_1 has no last element or Y_2 has no first element. Because $(X, \prec)$ is a *semiorder* and $(X, \prec)$ is connected, this leads to a contradiction: by connectedness there must be a finite path in $\sim$ between $y_1 \in Y_1$ and $y_2 \in Y_2$; hence if Y_1 has no last element or Y_2 has no first, then some interval in this path has an infinite number of Y_1 or Y_2 intervals in its interior, which contradicts (2.2) or the form of F.

Therefore, every $y \in Y$ that is not at an end of the chain has an immediate predecessor in Y and an immediate successor in Y, and every partition $Y_1 \prec Y_2$ of Y has a last element in Y_1 and a first element in Y_2. It then follows from standard results (e.g., Rosenstein, 1982, Chapter 3), that either Y is finite or is isomorphic to one of $(\{0, 1, 2, \ldots\}, <)$, $(\{\ldots, -1, 0\}, <)$, or $(\mathbb{Z}, <)$.

For simplicity, I shall assume that $(Y, <)$ has no first or last element, hence that it is isomorphic to $(\mathbb{Z}, <)$. The end points for the other cases are handled by slight modifications, as in Rabinovitch (1978a).

Given $(Y, \prec) \cong (\mathbb{Z}, <)$, we form a doubly-infinite partition $\{\ldots, C_{-2}, C_{-1}, C_0, C_1, C_2, \ldots\}$ of X in which some of the C_i can be empty. Let $C_0 = \{x_0\}$ for a fixed $x_0 \in X$. Let C_1 consist of all x whose intervals begin to the right of $\inf F(x_0)$ and have a nonempty intersection with $F(x_0)$. Let C_{-1} consist of all x whose intervals end before $\sup F(x_0)$ and intersect $F(x_0)$. Clearly, C_1 and C_{-1} are nonempty and disjoint.

Our general definition of C_{i+1} for $i > 0$ is:

if $C_i \neq \emptyset$, let C_{i+1} consist of all x (if any) whose intervals begin after those for C_i and intersect *all* the C_i intervals;

if $C_i = \emptyset$, let C_{i+1} consist of all x whose intervals begin after those for $C_0, \ldots, C_{i-1}$ and which intersect *at least one* of those intervals.

The general definition of C_{i-1} for $i < 0$ is symmetric, in the opposite direction away from C_0. Taking $A \sim B$ and $A \prec B$ to be true by definition when at

Dimensions of Semiorders

least one of A and B is empty, it is easily seen that $C_i \sim C_i$ for all i and that

(a) $\cdots \sim C_{-2} \sim C_{-1} \sim C_0 \sim C_1 \sim C_2 \sim \cdots$.
(b) If $C_i = \emptyset$ then $C_{i-1} \neq \emptyset$ and $C_{i+1} \neq \emptyset$.
(c) $C_i \prec C_j$ whenever $j - i \geq 3$.
(d) $\cup C_i = X$, with $C_i \cap C_j = \emptyset$ if $i \neq j$.

The construction obviously gives disjoint C_i that satisfy (a). It continues with nonempty C_{i+1} (in the positive direction) when $C_i = \emptyset$, by connectedness, and the definitions and the uniform character of F ensure that $C_i \sim C_i$ for all i—in particular for those whose predecessors in the definition process are empty. Conclusion (c) is straightforward to verify, and $\cup C_i$ covers X since otherwise, in view of (b) and (c), we would have to conclude that some maximal chain of $(X, \prec)$ is not isomorphic to $(\mathbb{Z}, <)$, which is impossible for the case under consideration.

The only relationships left open between the C_i concern C_i versus C_{i+2}. If $x \in C_i$ and $y \in C_{i+2}$, we could have either $x \sim y$ or $x \prec y$. This is where the usefulness of $[C_{i+2}|C_i]$ from Theorem 13 becomes apparent, for if $x \sim y$ in the case at hand, we need y less than x in some linear extension in a set that realizes $(X, \prec)$.

Three linear extensions of $(X, \prec)$ that realize $(X, \prec)$ are:

$$\prec_1 = \cdots [C_{-4}|C_{-6}][C_{-1}|C_{-3}][C_0|C_{-2}][C_3|C_1][C_4|C_2] \cdots$$
$$\prec_2 = \cdots [C_{-2}|C_{-4}][C_1|C_{-1}][C_2|C_0][C_5|C_3][C_6|C_4] \cdots$$
$$\prec_3 = \cdots d(C_{-2})d(C_{-3})d(C_0)d(C_{-1})d(C_2)d(C_1)d(C_4)d(C_3) \cdots$$

where $d(C_i)$ in $\prec_3$ is the dual of $\prec_2$ on C_i. Figure 5.3 illustrates the

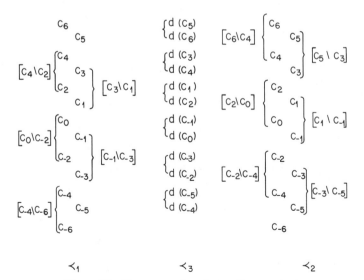

Figure 5.3 Linear extensions that realize a semiorder.

Figure 5.4 Irreducible semiorders of $H = 2$, $D = 3$.

formation of these linear extensions from the C_i, arranged by odd and even subscripts, and by the usual vertical poset diagram scheme. It is easily seen from the figure that $\prec \subseteq (\prec_1 \cap \prec_2 \cap \prec_3)$, and that if $x \sim y$ and $x \neq y$, then $x \prec_i y$ and $y \prec_j x$ for some $i, j \in \{1, 2, 3\}$. □

Rabinovitch (1978a) also shows that there are precisely three irreducible semiorders of height 2 that have dimension 3. One of these is IV in Fig. 5.1. The other two are shown in Fig. 5.4. Consequently, a height-2 semiorder has dimension 3 if and only if it has a seven-point restriction that is isomorphic to one of these three irreducible posets.

5.5 DIMENSIONS OF INTERVAL ORDERS

This section proves two basic results for dimensions of interval orders that augment the special cases in Theorems 10, 11, and 13. The first, due to Rabinovitch (1978b), is an extension of Theorem 10 that bounds $D(X, \prec)$ above by a function of the height of $(X, \prec)$. The second, from Bogart, Rabinovitch, and Trotter (1976), shows that there are finite interval orders of arbitrarily large dimension, hence, by Theorem 3(a), that there are interval orders of dimension n for every $n \geq 1$. The latter result involves an interesting application of Ramsey theory and ties in to magnitudes of interval orders.

In the following theorem, $\lceil x \rceil$ denotes the smallest integer greater than or equal to x.

Theorem 15 (Rabinovitch). *Suppose $(X, \prec)$ is an interval order of finite height. Then*

$$D(X, \prec) \leq \lceil \log_2(H(X, \prec) + 1) \rceil + 1.$$

Proof. Since an interval order with height strictly between $2^{n-1} - 1$ and $2^n - 1$ can obviously be embedded in an interval order of height $2^n - 1$ that has the same dimension, it will suffice to prove that, for finite $n \geq 2$,

$$H(X, \prec) = 2^n - 1 \Rightarrow D(X, \prec) \leq n + 1.$$

Assume henceforth that $(X, \prec)$ is an interval order of height $2^n - 1$, and, with no loss of generality, assume that $x \neq y \Rightarrow x \not\sim y$ so that $<^-$ and $<^+$ of the preceding section are linear orders on X.

For each $x \in X$ let $h(x)$ be the height of a maximum chain whose top (last) element is x, $0 \le h(x) \le 2^n - 1$. Also let

$$X(i) = \{x: h(x) = i\}, \quad 0 \le i \le 2^n - 1,$$

$$X(i, j) = \{x: i \le h(x) \le j\}, \quad 0 \le i \le j \le 2^n - 1.$$

We construct $\{\prec_1, \ldots, \prec_n, \prec_{n+1}\}$ that realizes $(X, \prec)$. For $i \le n$, $\prec_i$ splits the $X(j)$ into 2^{n-i} equal parts and, within each part, takes $[A|B]$ by an equal split of that part with larger indices in A:

$$\prec_1 = [X(1)|X(0)][X(3)|X(2)] \cdots [X(2^n - 1)|X(2^n - 2)]$$

$$\prec_2 = [X(2,3)|X(0,1)][X(6,7)|X(4,5)] \cdots$$

$$[X(2^n - 2, 2^n - 1)|X(2^n - 4, 2^n - 3)]$$

$$\prec_3 = [X(4,7)|X(0,3)][X(12,15)|X(8,11)] \cdots$$

$$[X(2^n - 4, 2^n - 1)|X(2^n - 8, 2^n - 5)]$$

$$\vdots$$

$$\prec_n = [X(2^{n-1}, 2^n - 1)|X(0, 2^{n-1} - 1)].$$

The final linear extension of $\prec$ is

$$\prec_{n+1} = d(X(0))d(X(1))d(X(2)) \cdots d(X(2^n - 1)),$$

where the dual is taken on the order of $X(j)$ in $\prec_1$. If $x \prec y$, so $h(x) < h(y)$, it is easily seen that x precedes y in all $\prec_i$. Since $X(i) \sim X(i)$, $\prec_1$ and $\prec_{n+1}$ take care of within-$X(i)$ cases. Finally, if $x \sim y$ when $h(x) < h(y)$, then $x \prec_{n+1} y$ and $y \prec_i x$ for some $i \le n$. □

In the rest of this section we shall focus on magnitude-m interval orders, with no equivalence between distinct points, whose characteristic matrices as defined in Section 2.3 have 1's in all cells. We shall refer to a magnitude-m interval order of this type as ψ_m. A specific example of $\psi_m = (X, \prec)$ is given by

$$X = \{[i, j]: 1 \le i \le j \le m, i \text{ and } j \text{ integers}\},$$

$$[i, j] \prec [i', j'] \Leftrightarrow j < i'.$$

Some parameters of ψ_m for $m \geq 2$ are

$$H(\psi_m) = m - 1,$$

$$W(\psi_m) = \left(\frac{m+1}{2}\right)^2 \text{ for odd } m,$$

$$W(\psi_m) = \left(\frac{m}{2}\right)\left(\frac{m}{2} + 1\right) \text{ for even } m,$$

$$B(\psi_m) = 2,$$

$$\text{depth}(\psi_m) = \left\lfloor \frac{m+1}{2} \right\rfloor,$$

where $\lfloor x \rfloor$ denotes the largest integer less than or equal to x. Verification of these values is left to the reader. They are illustrated (except for breadth) for $m = 7$ in Fig. 5.5, which shows the characteristic matrix of ψ_7. In the next section we shall note that $B(X, \prec) \leq 2$ for every interval order.

The dimension of ψ_m is immeasurably more difficult to determine. Bogart, Rabinovitch, and Trotter (1976) show that

$$D(\psi_2) = D(\psi_3) = 2$$
$$D(\psi_4) = \cdots = D(\psi_{11}) = 3$$
$$D(\psi_{12}) = 4,$$

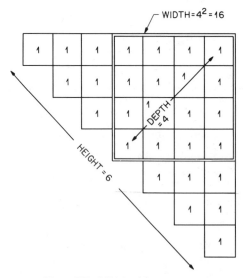

Figure 5.5 $M(\psi_7)$ with parameters.

and prove that

$$D(\psi_m) < \log_a(m-1) + \tfrac{1}{2}$$

where $a = 1 + 2^{1/2}$. Since $a > 2$, this is a much better bound on $D(\psi_m)$ for large m than that provided by Theorem 15.

The proof that $D(\psi_m)$ can be arbitrarily large uses the following version of Ramsey's theorem. A general account of Ramsey theory is given by Graham, Rothschild, and Spencer (1980).

Lemma 3. *Suppose $i, j, k \in \{1, 2, \ldots\}$. Then there is an integer m_0 such that if $|A| = m > m_0$ and the k-element subsets of A are partitioned into j (or fewer) parts, then there is an i-element subset B of A such that all k-element subsets of B are included in one of the j (or fewer) parts.*

Theorem 16 (Bogart, Rabinovitch, Trotter). *For every positive integer j there is an integer $m(j)$ such that $D(\psi_m) > j$ for all $m > m(j)$.*

Proof. Let $\psi_m = (X, \prec)$ with $X = \{[a, b]: 1 \le a \le b \le m,\ a$ and b integers$\}$ and $[a, b] \prec [c, d] \Leftrightarrow b < c$. Suppose ψ_m is realized by $\{\prec_1, \ldots, \prec_j\}$. Partition the three-element subsets $\{a < b < c\}$ of $\{1, \ldots, m\}$ into j or fewer classes by placing $\{a < b < c\}$ in class α if α is the smallest index for which $[b, c] \prec_\alpha [a, b]$.

To apply Lemma 3, take $i = 4$, j as just given, and $k = 3$. Then, when $m > m(j)$, Lemma 3 says that there is a four-element subset $\{a < b < c < d\}$ of $\{1, \ldots, m\}$ all of whose three-elements subsets lie in the same class, say class α. Then, by the formation of classes, $[c, d] \prec_\alpha [b, c]$ and $[b, c] \prec_\alpha [a, b]$, so $[c, d] \prec_\alpha [a, b]$ by transitivity. But then $\prec_\alpha$ is not a linear extension of $\prec$ since $[a, b] \prec [c, d]$. Hence, for given j, ψ_m cannot be realized by j linear extensions of ψ_m if m is sufficiently large. □

An interesting open problem posed by Bogart, Rabinovitch, and Trotter (1976) is to determine $D(\psi_m)$ precisely for all m.

5.6 BREADTH AND DEPTH

We note here a few more facts about breadth and depth for interval orders. The comments on depth in the latter part of the section provide a point of departure for our discussion in Chapters 9 and 10 of the number of lengths needed to represent a finite interval order or interval graph.

It has already been noted in Theorems 1 and 8 that $B(X, \prec) \le D(X, \prec)$ for a poset $(X, \prec)$ and that $B(X, \prec) = D(X, \prec)$ for some posets with large dimensions. The following theorem, from Baker, Fishburn, and Roberts (1970), shows that interval orders cannot be very broad. An auxiliary observation on depth (proof left to the reader) is included for semiorders.

Theorem 17 (Baker, Fishburn, and Roberts). *Suppose $(X, \prec)$ is an interval order. Then $B(X, \prec) \leq 2$. Moreover, $(X, \prec)$ is a semiorder if and only if depth $(X, \prec) = 1$.*

Proof. Suppose $B(X, \prec) > 2$. Then X includes a join-independent subset Y with more than two points, and any three-point subset of Y provides the lower three points of a copy of poset I in Fig. 5.1. However, that poset is not an interval order and, since every restriction of an interval order is an interval order, $(X, \prec)$ cannot be an interval order. □

Here, and later, we shall let $\rho^*(X, \prec)$ denote the fewest lengths needed to represent a finite interval order $(X, \prec)$ with real intervals. Formally,

$$\rho^*(X, \prec) = \min_{(f,\rho)} \{|\rho(X)|: (f,\rho) \text{ is a representation of } (X, \prec) \text{ with } \rho > 0\}.$$

The paragraph preceding Section 2.5 elaborates on the meaning of (f, ρ). Although $\rho \geq 0$ is specified there, it is easily seen that there is no loss of generality in requiring the length function ρ to be strictly positive for finite interval orders.

The definition of $\subset_0$ in Section 5.1 shows that if (f, ρ) is a representation of interval order $(X, \prec)$, and if $x \subset_0 y$, then $\rho(x) < \rho(y)$. Consequently, for all finite interval orders,

$$\rho^*(X, \prec) \geq \text{depth}(X, \prec).$$

If equality always held here, there would be little more to tell. However, that is not the case. In fact, as will be demonstrated in the latter part of Chapter 9, there are depth-2 interval orders with arbitrarily large values of ρ^*. This is in sharp contrast to $\rho^*(X, \prec) = 1$ for every depth-1 interval order, that is, for every semiorder: see Theorem 2.9.

On the other hand, equality often holds in the preceding expression. One important special case is noted by

Theorem 18. *Suppose $k \in \{2, 3, \ldots\}$ and $(X, \prec)$ is a depth-k interval order on $3k - 2$ points. Then $\rho^*(X, \prec) = k$.*

Proof. Figure 5.6 pictures a representation of a depth-k interval order on $3k - 2$ points, with $x_1 \subset_0 x_2 \subset_0 \cdots \subset_0 x_k$. Clearly $\rho^*(X, \prec) \geq k$ for every such interval order. Moreover, every depth-k interval order on $3k - 2$ points has a representation which looks like Fig. 5.6 in part. The only thing left open is the endpoints of the intervals, like a and b, which are not completely specified.

Clearly $\rho^*(X, \prec) = 2$ for $k = 2$. Assume then that $k \geq 3$ and that the theorem is true for smaller depths. Let $(X, \prec)$ be a depth-k interval order on $3k - 2$ points, and let $(Y, \prec)$ be its restriction on $3k - 5$ points when

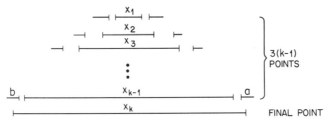

Figure 5.6 A depth-k interval order.

$\{a, b, x_k\}$ is removed from X (see the figure) to yield Y. By the induction hypothesis, $(Y, \prec)$ has a $(k-1)$-lengths representation, say (f', ρ'). We add $\{a, b, x_k\}$ back in to obtain a k-lengths representation for $(X, \prec)$ as follows. Position a, b, and x_k with a new longest length in the context of (f', ρ') so that a and b intersect x, a lies completely to the right of all $y \in Y$ for which $y \prec a$, and b lies completely to the left of all $y \in Y$ for which $b \prec y$. Finally, if other intervals for Y not in $\{x_1, \ldots, x_{k-1}\}$ must intersect a or b, extend them in the indicated direction using the new long length. A suitably long kth length for x, a, b, and perhaps other intervals as just indicated must give a k-lengths representation of $(X, \prec)$. □

5.7 NUMBERS OF RELATED SETS

We conclude this chapter with comments on numbers of interval orders and other binary relations on finite sets. Ordering relations will be discussed first, followed by graphs.

As usual, we distinguish between labeled and unlabeled counts. Two related sets that are defined on the same set X but are not identical will each be counted in the labeled case, but will count as only one related set in the unlabeled case if they are isomorphic. To be specific, let $R_n(\lambda)$ denote the set of all λ on $\{1, \ldots, n\}$. Thus, for example,

R_4(semiorders) is the set of all semiorders on $\{1, 2, 3, 4\}$.

Then let

$$N_n(\lambda) = |R_n(\lambda)| \quad \text{(labeled count)}$$

$$N_n(\lambda/\cong) = |R_n(\lambda)/\cong| \quad \text{(unlabeled count)}.$$

The difference between the labeled and unlabeled cases is obvious for linearly ordered sets:

$$N_n(\text{linear orders}) = n!$$

$$N_n(\text{linear orders}/\cong) = 1.$$

The situation for weak orders is a bit more complex:

$$N_n(\text{weak orders}/\cong) = 2^{n-1},$$

$$N_n(\text{weak orders}) = \sum_{j=1}^{n} \binom{n}{j} N_{n-j}(\text{weak orders}), \ [N_0 = 1],$$

$$= \sum_{k=1}^{n} \sum_{j=1}^{k} (-1)^{k-j} \binom{k}{j} j^n$$

$$= \sum_{v=0}^{\infty} v^n 2^{-(v+1)}.$$

The count under isomorphism arises from placements of single vertical bars between some or none of the pairs of points adjacent in $S = 1, 2, \ldots, n$, with each segment of S thus delineated interpreted as an equivalence class in a weak order and with these classes ordered left to right. Since $n - 1$ positions are available for vertical bars, $N_n(\text{weak orders}/\cong) = 2^{n-1}$.

The labeled count for weak orders traces back at least to Cayley (1859), as noted by Mendelson (1982), and has been rediscovered several times since. A very informative treatment is given by Gross (1962), who may have been the first to discuss the problem in terms of preference orderings. The recursive formula, $N_n = \Sigma \binom{n}{j} N_{n-j}$, follows from the observation that the first equivalence class can be formed in $\binom{n}{j}$ labeled ways from $\{1, \ldots, n\}$ when it contains j points, and then the remaining $n - j$ points have N_{n-j} ways to complete the weak order (with $N_0 = 1$). The double summation form for N_n can be verified by observing that it is the number of ways to place n labeled balls into k ordered cells with all cells occupied, summed over k from 1 to n. The final form arises from the fact that the exponential generating function for the N_n, that is, $\Sigma_{n=0}^{\infty}(N_n/n!)x^n$, equals $1/(2 - e^x)$: see, for example, Gross (1962) or Mendelson (1982).

As far as I am aware, a simple formula for semiorder or interval order counts is known only for unlabeled semiorders and is due to Wine and Freund (1957). See also Dean and Keller (1968).

Theorem 19 (Wine and Freund). *For each* $n \in \{1, 2, \ldots\}$,

$$N_n(\text{semiorders}/\cong) = \frac{1}{n+1}\binom{2n}{n}.$$

Proof. Let $X = \{x_1, x_2, \ldots, x_n\}$ and view $S = 12 \ldots n$ as the linearly ordered placements of the left endpoints of intervals for the x_i in a semiorder representation as in Theorem 2.7. (Point i in S corresponds to x_i, and no generality is lost by taking all endpoints distinct.) Let i' be a right endpoint designator for x_i. A semiorder, up to isomorphism, is obtained by inserting $1', 2', \ldots, n'$ into S *in that order* such that, for all i, i' is to the right of i in the joint linear arrangement of $\{1, \ldots, n\} \cup \{1', \ldots, n'\}$. The semiorder thus generated has $x_i \prec x_j$ if and only if i' precedes j (so $i < j$).

It should be clear that all semiorders on n points (up to isomorphism) are obtained in this way. Moreover, the semiorders generated by different insertions are not isomorphic as is easily seen by considering the smallest i' that is positioned differently in two cases. Consequently, $N_n(\text{semiorders}/\cong)$ is the number of linear orders on $\{1, \ldots, n\} \cup \{1', \ldots, n'\}$ that have restrictions $12 \ldots n$ and $1'2' \ldots n'$ and have i before i' for all i.

Let $f_n(j)$ be the number of such linear orders that have $1'$ immediately after j for $j = 1, \ldots, n$. Also let $f_n(0) = 0$ for all n and $f_n(n+1) = 0$ for all n. It is easily seen that, by adding one interval beginning on the far left (use 0 and $0'$ say), we have

$$f_{n+1}(k) = \sum_{j=k-1}^{n} f_n(j), \qquad k = 1, \ldots, n+1.$$

Consequently,

$$f_{n+1}(k) = f_{n+1}(k+1) + f_n(k-1), \qquad n \geq 1, \ 1 \leq k \leq n+1.$$

Since $f_1(1) = 1$, a quick check shows that the solution to these equations is

$$f_n(j) = \frac{j}{2n-j} \binom{2n-j}{n-j}.$$

Moreover, because $N_n(\text{semiorders}/\cong)$ equals $\sum_j f_n(j)$ by definition and, for $n \geq 1$, $f_{n+1}(1) = \sum_j f_n(j)$, we have

$$N_n(\text{semiorders}/\cong) = f_{n+1}(1) = \frac{1}{2n+1}\binom{2n+1}{n}$$

$$= \frac{1}{n+1}\binom{2n}{n}. \qquad \square$$

The method of the preceding proof can be generalized to count unlabeled interval orders on n points. In view of Theorem 2.5, a magnitude-m interval order on n points is uniquely identified up to isomorphism by

$$(A_1, B_1, A_2, B_2, \ldots, A_m, B_m)$$

where $\{A_1, \ldots, A_m\}$ is a partition of $\{1, 2, \ldots, n\}$ with all A_j nonempty and $A_j < A_{j+1}$ (all i in A_j are smaller than all i in A_{j+1} for the unlabeled case); $\{B_1, \ldots, B_m\}$ is a partition of $\{1', 2', \ldots, n'\}$ with all B_j nonempty; if i_1 and i_2 are in the same A_j with $i_1 < i_2$, then i_2' does not appear in an earlier B_k than i_1'

(to avoid multiple counting in the unlabeled case); and, for all i, if $i \in A_j$ then $i' \in B_k$ for $k \geq j$. If we let $f_{n,m}(k)$ be the number of unlabeled magnitude-m interval orders on n points for which $|A_1| = k$ for $k = 1, \ldots, n - m + 1$, and let $f_{n,m} = N_n$(interval orders of magnitude $m/\cong$), then

$$f_{n,m} = \sum_{k=1}^{n-m+1} f_{n,m}(k), \qquad N_n(\text{interval orders}/\cong) = \sum_{m=1}^{n} f_{n,m}.$$

Using the characterization of nonisomorphic interval orders given here, it is easily seen that $f_{n,1} = f_{n,n} = 1$, $f_{n,2}(k) = k$, $f_{n,2} = \binom{n}{2}$, and

$$f_{n,3}(k) = \binom{k+1}{2} f_{n-k,2} + \binom{k}{2}(n - k - 1),$$

$$f_{n,3} = \sum_{k=1}^{n-2} \left[\binom{k+1}{2}\binom{n-k}{2} + \binom{k}{2}(n - k - 1) \right].$$

Computation beyond $m = 3$ becomes cumbersome very quickly.

The lack of more definitive results for interval orders and labeled semiorders lies in sharp contrast to Hanlon (1982), which gives a thorough treatment of counts for interval graphs and indifference graphs. Although Hanlon's generating-function approach would appear to be adaptable to interval orders and semiorders (which he uses as intermediaries to count graphs), I am not aware that this has been done.

Among other things, Hanlon derives generating functions for what I have called $N_n(\lambda)$ and $N_n(\lambda/\cong)$ for $\lambda \in$ {indifference graphs (= unit interval graphs), interval graphs, connected indifference graphs, connected interval graphs}, then uses these to compute specific numerical results. Extensive tables are provided (some n's are mislabeled in Table VII) along with asymptotic estimates for types of indifference graphs.

6

Embedded Semiorders and Indifference Graphs

This chapter is the first of two devoted to extremization problems for finite interval orders and interval graphs. It is concerned with maximum semiordered restrictions of interval orders, or equivalently, with maximum indifference subgraphs of interval graphs. Its specific focus is the largest integer k such that every n-point interval order includes a k-point semiorder. We also consider here a similar problem when the n-point interval orders have no antichain with more than j points. The other extremization chapter, Chapter 10, looks at the problem of minimizing the number of lengths in real representations of interval orders and interval graphs.

6.1 EXTREMIZATION PROBLEMS

The theme of this chapter is a very familiar one in the theory of related sets, namely to determine the largest or smallest configuration, or integer for the size of a largest or smallest configuration, of a specified type, perhaps in relation to configurations of another type. Two prominent examples illustrate this theme.

The first concerns asymmetric and complete related sets, commonly known as *tournaments*. What is the largest integer k such that every tournament on n points includes a k-point chain? Or, given k, what is the smallest n such that every tournament on n or more points includes a k-point chain? You can quickly check that the answer to the latter question is $n = 4$, given $k = 3$, but it was not known until about 1968 that the answer for $k = 5$ is $n = 14$ (Reid and Parker, 1970). A dated but still very useful introduction to tournaments is provided by Moon (1968).

The second prominent example comes from Ramsey theory. Given positive integers j and k, what is the smallest n such that every irreflexive graph on n points has either a j-point clique or a k-point independent set? It is easily seen that $n = 6$ for $(j, k) = (3, 3)$, and Gleason and Greenwood (1955) proved that $n(3, 4) = 9$, $n(3, 5) = 14$, and $n(4, 4) = 18$. Only a few other values for $(j, k) \geq (3, 3)$ are presently known despite huge efforts, and even $n(5, 5)$ seems well beyond determination by present capabilities. This and other facets of Ramsey theory are discussed at length in Graham, Rothschild, and Spencer (1980).

Our present concern with maximum semiorders in interval orders, or maximum indifference graphs in interval graphs, is illustrated by the 12-point interval order and its associated interval graph in Fig 6.1. By deleting the long intervals numbered 4, 5, and 8, which prevent the interval order from being a semiorder, we obtain the semiorder and its associated indifference graph shown in the lower part of the figure. However, this is *not* a maximum semiorder included in the interval order, for there is exactly one 10-point semiorder in the interval order. It is obtained by deleting interval 5 and one other interval. Which one?

Our investigation of maximum semiorders in interval orders will focus on the function τ on $\{1, 2, \ldots\}$ defined by

$$\tau(n) = \max\{k: \text{every } n\text{-point interval order has a } k\text{-point semiorder restriction}\}.$$

Clearly, τ is nondecreasing in n and, if it is known, then for each k we know the smallest n such that every n-point interval order includes a k-point semiorder.

We shall refer to a four-point subset of an interval order that violates property (2.2) of Section 2.1 as a $K_{1,3}^*$ set. A typical $K_{1,3}^*$ set is $\{a \prec b \prec c, x \sim \{a, b, c\}\}$. Since a restriction of an interval order is a semiorder if and only

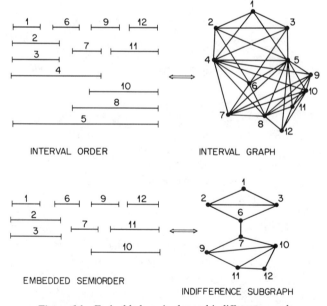

Figure 6.1 Embedded semiorder and indifference graph.

Extremization Problems

if the restriction has no $K_{1,3}^*$ set, $\tau(n)$ is the largest k such that every n-point interval order has a k-point restriction with no $K_{1,3}^*$ set. Moreover, because a subgraph of an interval graph is an indifference graph if and only if the subgraph has no copy of $K_{1,3}$ (Theorem 3.3), and there is a 1-1 correspondence between the $K_{1,3}$ subgraphs of an interval graph and the $K_{1,3}^*$ sets of every interval order that agrees with the interval graph, it follows that

$\tau(n) = \max\{k$: every n-point interval graph has a k-point indifference subgraph$\}$

$= \max\{k$: every n-point interval graph has a k-point subgraph with no copy of $K_{1,3}\}$.

Consequently, the results for τ apply to indifference graphs within interval graphs as well as to semiorders within interval orders.

It is obvious that $\tau(1) = 1$, $\tau(2) = 2$ and $\tau(3) = \tau(4) = 3$. The next section shows that $\tau(n)$ for even n is bounded above by $(n/2) + 1$ and that this bound is tight for small even n. The third section derives a much smaller upper bound for large n and shows also that

$$\tau(n) \geq \frac{n}{\log_2 n} \quad \text{for} \quad n \geq 3.$$

Although it seems likely that $\tau(n)(\log_2 n)/n$ converges to a constant in $[1, 3]$, the question of convergence remains open.

The final section of the chapter examines

$\tau_j(n) = \max\{k$: every n-point interval order with no antichain of more than j points includes a k-point semiorder$\}$

for $j = 2, 3$. Clearly,

$$n = \tau_1(n) \geq \tau_2(n) \geq \tau_3(n) \geq \cdots \geq \tau_n(n) = \tau(n).$$

We shall see that τ_2 and τ_3 have repeating patterns of lengths 4 and 5 respectively.

Our proofs will use two facts that will not always be noted explicitly. First, since two antichains cannot include a $K_{1,3}^*$ set, the largest semiorder in an interval order must contain at least as many points as appear in any two disjoint antichains. Second, if $X_1, X_2, \ldots, X_N$ are subsets of X with $X_1 \prec X_2 \prec \cdots \prec X_N$, then the largest semiorder in $(X, \prec)$ must have at least as many points as the sum of the cardinalities of the largest semiorders in $(X_1, \prec)$ through $(X_N, \prec)$.

6.2 UPPER BOUND AND EXACT VALUES

This section concentrates on $\tau(n)$ for small n. We shall prove two theorems.

Theorem 1. $\tau(n-1) \leq \tau(n) \leq (n/2)+1$ *for all even* $n \geq 4$, *and equality holds both places if* $n \in \{4, 6, \ldots, 14\}$.

Theorem 2. $\tau(15) = \tau(16) = \tau(17)$.

It follows that the $\tau(n)$ values for n from 1 through 17 are

$$1, 2, 3, 3, 4, 4, 5, 5, 6, 6, 7, 7, 8, 8, 9, 9, 9.$$

Hence the pattern of single repeats $(3, 3; 4, 4; \ldots)$ breaks down for $n > 14$. Alternatively, with $\tau^*(k)$ the smallest n such that every n-point interval order includes a k-point semiorder

$$\tau^*(k) = 2k - 3 \quad \text{for} \quad 3 \leq k \leq 9,$$

but $\tau^*(10) \geq 18 > 2(10) - 3 = 17$. The results in the next section imply that $\tau^*(k+1) - \tau^*(k)$ takes on arbitrarily large values.

The proofs of Theorems 1 and 2 will be divided among four lemmas.

Lemma 1. $\tau(n) \leq (n/2) + 1$ *for even* $n \geq 4$.

Lemma 2. $\tau(17) = 9$.

Lemma 3. $\tau(n) \geq (n+3)/2$ *for odd* $n \in \{3, \ldots, 13\}$.

Lemma 4. $\tau(15) \geq 9$.

Proof of Lemma 1. For even $n \geq 4$, let p_n be an n-point interval order consisting of an $[(n/2) + 1]$-point chain plus $(n/2) - 1$ points that are universal in the corresponding interval graph. A representation of p_n is pictured in Fig. 6.2(a). The maximum semiorders in p_n are clearly the $[(n/2) + 1]$-point chain and the $(n/2) + 1$ points in two antichains that include the $(n/2) - 1$ universal points along with two points in the chain. Hence $\tau(n) \leq (n/2) + 1$. □

Similar reasoning for Figs. 6.2(b) through 6.2(d) shows that $\tau(17) \leq 9$, $\tau(19) \leq 10$, and $\tau(21) \leq 11$. The dashed verticals identify two disjoint antichains that maximize the number of points in two such antichains.

Proof of Lemma 2. It remains to show that $\tau(17) \geq 9$. Let $X = \{x_1, \ldots, x_{17}\}$ with $|X| = 17$, and let $F_i = [F_i^-, F_i^+]$ be the interval for x_i in a

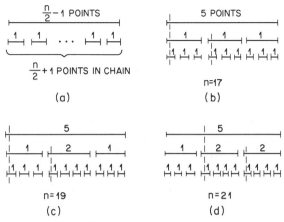

Figure 6.2 Interval order representations.

representation F of an interval order on X. Assume with no loss of generality that

$$F_1^- \leq F_2^- \leq \cdots \leq F_{17}^-.$$

Let h be the number of $i \leq 8$ for which $F_i^+ < F_9^-$. If $h \geq 3$, then a three-point semiorder from these h points plus a six-point semiorder from $\{x_9, \ldots, x_{17}\}$ (by $\tau(9) \geq 6$, which is left to the reader to prove) yield a nine-point semiorder overall. If $h \leq 1$, then $\{x_1, \ldots, x_9\}$ is a semiorder (one or two antichains).

Suppose henceforth that $h = 2$. Then $\{x_1, \ldots, x_9\}$ includes a seven-point antichain. If $\{x_{10}, \ldots, x_{17}\}$ has a two-point antichain, we are done. Otherwise, $\{x_{10}, \ldots, x_{17}\}$ is an eight-point chain, which yields a 10-point semiorder in conjunction with the two points that have $F_i^+ < F_9^-$. □

Proof of Lemma 3. Clearly, $\tau(3) = 3$. The proofs that $\tau(n) \geq (n + 3)/2$ for $n \in \{5, 7, 9\}$ are left to the reader. We proceed with $n = 11$, then consider $n = 13$.

For $\tau(11) \geq 7$, we adopt the notation of the preceding proof and assume for definiteness that $F_1^- \leq F_2^- \leq \cdots \leq F_{11}^-$. If $F_i^+ < F_7^-$ for at least three $i \leq 6$, then, since the five intervals for x_7 through x_{11} begin after these three or more end, the interval order includes a semiorder with at least $\tau(3) + \tau(5) = 3 + 4 = 7$ points.

To complete the proof that $\tau(11) \geq 7$, suppose at most two $i \leq 6$ have $F_i^+ < F_7^-$. Then at least five points in $\{x_1, \ldots, x_7\}$, including x_7, form an antichain. If $\{x_8, x_9, x_{10}, x_{11}\}$ includes a two-point antichain, we are done, so assume to the contrary that $x_8 \prec x_9 \prec x_{10} \prec x_{11}$. Then, if three or more $i \leq 7$ have $F_i^+ \leq F_8^-$, we obtain a semiorder with at least $\tau(3) + 4 = 7$ points. Otherwise, $\{x_1, \ldots, x_8\}$ includes a six-point antichain, which with any other point gives a seven-point semiorder. Therefore $\tau(11) \geq 7$.

Assume henceforth in this proof that $(\{x_1,\ldots,x_{13}\}, \prec)$ is a 13-point interval order with $F_1^- \leq \cdots \leq F_{13}^-$, and let $h = |\{i \leq 7: F_i^+ < F_8^-\}|$. We consider four exhaustive cases to prove that $\tau(13) \geq 8$.

CASE 1. $h \geq 5$. There is a semiorder included in the interval order that has $\tau(5) + \tau(6) = 8$ points.

CASE 2. $h \leq 2$. Then $\{x_1,\ldots,x_8\}$ includes a six-point antichain. Since we are done if $\{x_9,\ldots,x_{13}\}$ has a two-point antichain, suppose $x_9 \prec x_{10} \prec x_{11} \prec x_{12} \prec x_{13}$. If $F_i^+ < F_9^-$ for three or more $i \leq 8$, we get a semiorder with at least $\tau(3) + 5 = 8$ points; if $F_i^+ < F_9^-$ for fewer than three $i \leq 8$, then $\{x_1,\ldots,x_9\}$ includes a seven-point antichain and hence an eight-point semiorder.

CASE 3. $h = 3$. Then $\{x_1,\ldots,x_8\}$ has a five-point antichain that contains x_8. We shall suppose that there is no eight-point semiorder and derive a contradiction. Thus, assume henceforth for this case that $\{x_9,\ldots,x_{13}\}$ has no three-point antichain. Assume also that $F_i^+ < F_9^-$ for at most four $i \leq 8$, since otherwise we get a semiorder with $\tau(5) + \tau(5) = 8$ points.

Suppose exactly three $i \leq 8$ have $F_i^+ < F_9^-$. Then $\{x_1,\ldots,x_9\}$ has a six-point antichain. Hence, to preclude an eight-point semiorder, we need $x_{10} \prec x_{11} \prec x_{12} \prec x_{13}$. Given this four-point chain: if $F_i^+ < F_{10}^-$ for at least five $i \leq 9$, there is a semiorder with $\tau(5) + 4 = 8$ points; if $F_i^+ < F_{10}^-$ for fewer than four i, then $\{x_1,\ldots,x_{10}\}$ has a seven-point antichain, hence an eight-point semiorder; and if $F_i^+ < F_{10}^-$ for exactly four i, then $\{x_1,\ldots,x_{10}\}$ includes a six-point antichain, so the four x_i with $F_i^+ < F_{10}^-$ must form a chain to prevent an eight-point semiorder, in which case these four x_i plus x_{10} through x_{13} yield an eight-point chain.

Thus, our supposition that there is no eight-point semiorder implies that exactly four $i \leq 8$ have $F_i^+ \leq F_9^-$. These four consist of the set A of the three $i \leq 7$ that give $h = 3$, plus one other x_i for $i \leq 7$, which we denote as x_j. (If the fourth point were x_8 then $A \cup \{x_8\}$ would be a semiorder, which combines with a four-point semiorder from $\{x_9,\ldots,x_{13}\}$ to yield an eight-point semiorder.) Let

$$B = \{x_1,\ldots,x_7\} \setminus (A \cup \{x_j\}),$$

the other three points in $\{x_1,\ldots,x_7\}$, each of which stands in the relationship $\sim$ (symmetric complement of $\prec$) to x_8 and x_9. To prevent an eight-point semiorder, it is easily seen that $A \cup \{x_j\}$ must be a $K_{1,3}^*$ set, and all three points in B must bear $\sim$ to a point in A, say x_k, whose interval extends farthest to the right. Therefore $B \cup \{x_j, x_k\}$ is a five-point antichain, which forbids $\{x_8,\ldots,x_{13}\}$ from having a three-point antichain. Figure 6.3 indicates the interval diagram that applies at this point in our analysis of Case 3.

With respect to Fig. 6.3, suppose first that $x_9 \prec x_{10}$, that is, that $F_9^+ < F_{10}^-$. If $\{x_{10},\ldots,x_{13}\}$ forms a $K_{1,3}^*$ set, then F_8 can intersect at most F_{10} of F_{10}

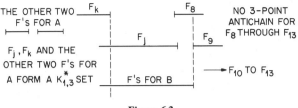

Figure 6.3

through F_{13}, and we get an eight-point semiorder for $A \cup \{x_8, x_9\} \cup \{x_{11}, x_{12}, x_{13}\}$; if $\{x_{10}, \ldots, x_{13}\}$ is not a $K_{1,3}^*$ set, then $A \cup \{x_9\} \cup \{x_{10}, x_{11}, x_{12}, x_{13}\}$ yields an eight-point semiorder. Suppose next that $x_9 \sim x_{10}$, so F_9 and F_{10} intersect. Then $F_8^+ < F_{10}^-$: if $x_9 \sim x_{11}$ then $x_{10} \prec x_{11}$ and $A \cup \{x_8, x_{10}, x_{11}, x_{12}, x_{13}\}$ gives an eight-point semiorder; if $x_9 \prec x_{11}$ then $A \cup \{x_8, x_9, x_{11}, x_{12}, x_{13}\}$ yields an eight-point semiorder. Since this exhausts the possibilities, we contradict the supposition for Case 3 that there is no eight-point semiorder.

CASE 4. $h = 4$. Let C denote the four x_i for which $F_i^+ < F_8^-$, and take $D = \{x_1, \ldots, x_7\} \setminus C$. If C is not a $K_{1,3}^*$ set, then C plus four from $\{x_8, \ldots, x_{13}\}$ yield an eight-point semiorder. Assume henceforth that C is a $K_{1,3}^*$ set, with x_j and x_k its points whose intervals extend farthest to the right. If at least one point in D does not bear $\sim$ to both x_j and x_k, then this point, along with x_8 though x_{13}, has a five-point semiorder ($\tau(7) = 5$) which combines with a three-point semiorder from C to give an eight-point semiorder. Assume therefore that all three points in D bear $\sim$ to both x_j and x_k, that is, $D \sim \{x_j, x_k\}$, so that these five points form an antichain. Then, if $\{x_8, \ldots, x_{13}\}$ has a three-point antichain, we are done. Otherwise, a diagram similar to Fig. 6.3 applies, and an argument like that used in the preceding paragraph shows that there must be an eight-point semiorder. □

Proof of Lemma 4. The proof that $\tau(15) \geq 9$ follows the same general method as the preceding proof. We presume that $(\{x_1, \ldots, x_{15}\}, \prec)$ is a 15-point interval order having a representation F with $F(x_i) = F_i = [F_i^-, F_i^+]$ and
$$F_1^- \leq F_2^- \leq \cdots \leq F_{15}^-.$$
In the present situation our cases are based on $h = |\{i: F_i^+ < F_9^-\}|$, the number of x_i whose intervals end before F_9 begins. For convenience, let $x_i^j = \{x_i, x_{i+1}, \ldots, x_j\}$.

CASE 1. $h \leq 1$. x_1^9 is a semiorder.

CASE 2. $h \geq 5$. Since $\tau(5) = 4$, the five or more x_i with $F_i^+ < F_9^-$ include a four-point semiorder. Since $\tau(7) = 5$, x_9^{15} has a five-point semiorder. The combination gives a nine-point semiorder.

CASE 3. $h = 2$. Let F_i and F_j lie completely to the left of F_9. Since x_1^9 has a seven-point antichain, we get a nine-point semiorder if $\{x_i, x_j\} \cup x_{10}^{15}$ has a two-point antichain. Otherwise, $\{x_i, x_j\} \cup x_{10}^{15}$ is an eight-point chain, say $x_i \prec x_j \prec x_{10} \prec \cdots \prec x_{15}$. These eight plus one from $A = x_1^{15} \setminus \{x_i, x_j, x_{10}, \ldots, x_{15}\}$ give a nine-point semiorder unless every $x \in A$ forms a $K_{1,3}^*$ set with three points in the eight-point chain. But then, since $|A| = 7$, $A \cup \{x_j, x_{10}\}$ yields a nine-point semiorder.

CASE 4. $h = 3$. We suppose there is no nine-point semiorder and obtain a contradiction. First, reindex the first eight x_i so that

$$F_1^+ \leq F_2^+ \leq F_3^+ < F_9^- \leq F_4^+ \leq \cdots \leq F_8^+.$$

We no longer assume that $F_4^- \leq \cdots \leq F_8^-$ although $F_i^- \leq F_9^-$ for every $i < 9$.

Given this reindexing, x_4^9 is a six-point antichain. To avoid a nine-point semiorder, assume henceforth for Case 4 that

(a) x_{10}^{15} has no three-point antichain.

Then, as is easily verified (or see Theorem 5 in Section 6.4), x_{10}^{15} has a five-point semiorder.

Suppose x_{10} is in an antichain with x_4 and x_9. Then x_4^{10} is a seven-point antichain, so to avoid a two-point antichain elsewhere we require $x_1^3 \cup x_{11}^{15}$ to be an eight-point chain. But then, as for Case 3, either this chain plus one point in x_4^{10} give a nine-point semiorder, or x_4^{10} plus two points in the chain yield a nine-point semiorder. Thus, to avoid a nine-point semiorder, x_{10} must not be in an antichain with x_4 and x_9: either $F_4^+ < F_{10}^-$ or $F_9^+ < F_{10}^-$. However, if $F_9^+ < F_{10}^-$, then $\{x_1, x_2, x_3, x_9\}$ plus a five-point semiorder from x_{10}^{15}, via (a), gives a nine-point semiorder. Therefore, assume henceforth that

(b) $F_4^+ < F_{10}^- \leq F_9^+$.

Suppose next that $F_5^+ < F_{10}^-$. Then, since $F_4^+ < F_{10}^-$ by (b), a nine-point semiorder results from a four-point semiorder in x_1^5 and a five-point semiorder in x_{10}^{15}. To avoid this, assume henceforth that

(c) $F_{10}^- \leq F_5^+$.

Suppose x_9^{15} has no three-point antichain. Then it is not hard to show that this seven-point set includes a six-point semiorder (see Theorem 5) which, in union with x_1^3, yields a nine-point semiorder. In view of this and (a), assume henceforth that

(d) x_9 and two x_i in x_{10}^{15} form a three-point antichain.

We now consider smaller i. Since $F_4^+ < F_{10}^-$ by (b), we get a nine-point semiorder with five points from x_{10}^{15}, plus x_1^4, unless x_1^4 is a $K_{1,3}^*$ set, which requires

(e) $F_4^- \leq F_3^+$.

Moreover, to avoid a nine-point semiorder, we need

(f) $F_i^- \leq F_3^+$ for exactly four $i \in \{4, 5, 6, 7, 8\}$.

For, if two or more F_i for $4 \leq i \leq 8$ are to the right of F_3, then these plus x_9^{15} yield a six-point semiorder ($\tau(9) = 6$), which in union with x_1^3 gives a nine-point semiorder; and, if all five of F_4 through F_8 intersect F_3, then x_3^8 is a six-point antichain whose union with a three-point antichain in x_9^{15}, via (d), gives a nine-point semiorder.

Henceforth, let j denote the point in $\{4, \ldots, 8\}$ for which $F_3^+ < F_j^-$ according to (f). By (e), $j \geq 5$. Figure 6.4 pictures F_j along with the other intervals developed to this point. Each of (a) through (f) is depicted at least partly in the figure. We use i to denote F_i.

Continuing with Case 4, we note next that if both F_j and F_{10} end before F_{11}, then $x_1^3 \cup \{x_j, x_{10}\}$ plus four points from x_{11}^{15} form a nine-point semiorder, and if both F_j and F_{10} intersect F_{11} then $\{x_j\} \cup x_9^{11}$ and $x_3^8 \setminus \{x_j\}$ are disjoint antichains with nine points. Therefore, assume henceforth that either

$$F_j^+ < F_{11}^- \leq F_{10}^+,$$

or

$$F_{10}^+ < F_{11}^- \leq F_j^+.$$

We consider these in turn. Suppose first that $F_j^+ < F_{11}^- \leq F_{10}^+$. Since F_{10} intersects F_{11}, (a) requires $x_{10} \prec x_{12}$ or $x_{11} \prec x_{12}$. If $x_{10} \prec x_{12}$, then a nine-point semiorder arises from $x_1^3 \cup \{x_j\}$ and five points in x_{10}^{15}. (If x_{10} is in the latter five, it cannot be in a $K_{1,3}^*$ set along with x_j since $x_{10} \prec x_{12}$.) Assume henceforth in this paragraph that

$$F_{11}^+ < F_{12}^- \leq F_{10}^+.$$

If F_{12} ends before F_{13} begins, then $x_1^3 \cup \{x_j\} \cup x_{11}^{15}$ is a nine-point semiorder, so assume also that $F_{13}^- \leq F_{12}^+$. Then, using (a), $F_{10}^+ < F_{13}^-$; that is, $x_{10} \prec x_{13}$. But then $x_1^3 \cup \{x_j, x_{10}, x_{11}\} \cup x_{13}^{15}$ is a nine-point semiorder. Thus, (a) though (f) plus $F_j^+ < F_{11}^- \leq F_{10}^+$ force a nine-point semiorder.

Suppose next that $F_{10}^+ < F_{11}^- \leq F_j^+$. Then, since $x_{10} \prec x_{11}$, (d) implies that F_9 intersects F_{12}, or $F_{12}^- \leq F_9^+$. Now if F_{11} and F_j also intersect F_{12}, then

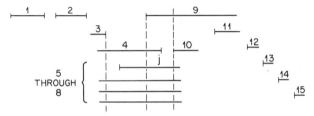

Figure 6.4 Trying to avoid a nine-point semiorder.

$x_3^{12} \setminus \{x_{10}\}$ equals two antichains and is therefore a nine-point semiorder. And if F_{11} and F_j end prior to F_{12}, then $x_1^3 \cup \{x_j, x_{10}, x_{11}\} \cup x_{13}^{15}$ is a nine-point semiorder. Moreover, the same semiorder arises if F_{11} and F_j are left of F_{13}. Hence, to avoid a nine-point semiorder, we must have either

$$x_j \prec x_{12} \quad \text{and} \quad F_{13}^- \leq F_{11}^+$$

or

$$x_{11} \prec x_{12} \quad \text{and} \quad F_{13}^- \leq F_j^+.$$

If the first of these holds, then (a) implies $x_{12} \prec x_{13}$; hence $x_1^3 \cup \{x_j, x_{10}, x_{12}\} \cup x_{13}^{15}$ is a nine-point semiorder. If the second holds, then (d) implies that F_9 intersects F_{13}. Then, if F_{12} also intersects F_{13}, the nine-point set $x_3^9 \cup x_{12}^{13}$ decomposes into two antichains; and, if $x_{12} \prec x_{13}$, then $x_1^3 \cup x_{10}^{15}$ is a nine-point semiorder. Hence, (a) through (f) plus $F_{10}^+ < F_{11}^- \leq F_j^+$ force a nine-point semiorder, and this completes the proof of Case 4.

CASE 5. $h = 4$. Suppose that, instead of the initial indexing of left endpoints, we begin with the ordering $F_1^+ \leq F_2^+ \leq \cdots \leq F_{15}^+$ of right endpoints and let $h' = |\{i: F_7^+ < F_i^-\}|$. Then the foregoing proofs for $h \neq 4$ apply symmetrically to $h' \neq 4$. Therefore, we need only consider the case in which $h = h' = 4$ to conclude the proof of Lemma 4.

Given $X = \{x_1, \ldots, x_{15}\}$ and $h = h' = 4$, let $\{x_1, x_2, x_3, x_4\}$ be the points for $h = 4$, and let $\{x_{12}, x_{13}, x_{14}, x_{15}\}$ be the points for $h' = 4$. If x_1^4 is not a $K_{1,3}^*$ set, then its four points plus five points from the seven-point set whose F_i are to the right of F_1 through F_4 form a nine-point semiorder. Since a symmetric result holds for x_{12}^{15}, assume henceforth that x_1^4 and x_{12}^{15} are $K_{1,3}^*$ sets with $x_1 \prec x_2 \prec x_3$ and $x_{13} \prec x_{14} \prec x_{15}$. Also let

$$J = \left[\min\{F_3^+, F_4^+\}, \max\{F_{12}^-, F_{13}^-\}\right].$$

If the intervals for three points in x_5^{11} lie wholly within J, then these three along with some three from each of x_1^4 and x_{12}^{15} yield a nine-point semiorder. On the other hand, if five of the seven points in x_5^{11} have intervals not wholly within J, these five plus x_3, x_4, x_{12}, and x_{13} constitute two antichains.

Therefore, $(X, \prec)$ has a nine-point semiorder when $h = h' = 4$, and the proof of Lemma 4 is complete. □

6.3 GENERAL BOUNDS

As n increases, it appears that the rate of increase of τ diminishes gradually. We have seen by Theorems 1 and 2 that $\tau(n)$ for small odd $n \geq 3$ grows linearly since $\tau(n) = (n + 3)/2$ for odd n from 3 to 15. However, at $n = 17$,

General Bounds

$\tau(n) < (n + 3)/2$. In the present section, where all logarithms are to base 2, we shall prove that

$$\frac{n}{\log n} \leq \tau(n) < \frac{7n}{\log n} \quad \text{for all } n \geq 3.$$

Moreover, if $\tau(n)(\log n)/n$ converges, its limiting value cannot exceed 3. The reason for 3, and not 7 as in the preceding inequality, is that we get an upper bound on $\tau(n)$ for a subsequence of n values of the form $n = 2^k(k+4)$ such that this bound times $(\log n)/n$ converges to 3.

We begin with the lower bound. The upper bound will be considered after the proof of the following theorem.

Theorem 3. $\tau(n) \geq n/(\log_2 n)$ for all $n \geq 3$.

Proof. The theorem follows from Theorem 1 for small n, to at least $n = 32$ since $32/(\log 32) = 6.4$. For larger n we shall let

$$h(r) = \frac{r}{\log r} \quad \text{and} \quad g(r) = \lceil h(r) \rceil \quad \text{for real } r > 0.$$

To prove that $\tau(n) \geq h(n)$ for $n > 32$, let $(X, \prec)$ be an interval order on n points with real representation F, and assume that the theorem holds for $n' < n$. If $(X, \prec)$ has two antichains with $g(n)$ or more points, we are done, so assume that no two antichains contain more than $g(n) - 1$ points.

For each real number r, let

$$f_1(r) = |\{x : \sup F(x) < r\}|$$
$$f_2(r) = |\{x : \inf F(x) > r\}|$$

so $f_1(r) + f_2(r) \leq n$ for all r. Moreover, by the two-antichains restriction, a jump in f_1 (as r increases) and a jump in f_2 (as r decreases) cannot jointly involve more than $g(n) - 1$ points in X. Let r_i be an r that minimizes $f_i(r)$ subject to

$$f_i(r) \geq \frac{n + 2 - 2g(n)}{3}$$

and let $n_i = f(r_i)$ for $i = 1, 2$. Because $n \geq 32$, it is easily seen that $r_1 < r_2$: otherwise some antichain has $g(n)$ or more points. Moreover, because the last jumps in the f_i before the r_i are reached cannot involve more than $g(n) - 1$ points,

$$n_1 + n_2 < \frac{2[n + 2 - 2g(n)]}{3} + g(n) - 1 = \frac{2n + 1 - g(n)}{3}.$$

Let n_3 be the number of points in X whose intervals lie strictly between r_1 and r_2. Since no more than $g(n) - 1$ intervals contain r_1 or r_2,

$$n_1 + n_2 + n_3 + g(n) - 1 \geq n,$$

and therefore

$$n_3 \geq n + 1 - g(n) - (n_1 + n_2)$$
$$> n + 1 - g(n) - \frac{2n + 1 - g(n)}{3} = \frac{n + 2 - 2g(n)}{3}.$$

Thus, $(X, \prec)$ includes a semiorder on at least $g(n_1) + g(n_2) + g(n_3)$ points, assuming of course that the theorem holds for $n' < n$, with

$$n_i \geq \frac{n + 2 - 2g(n)}{3} \quad \text{for} \quad i = 1, 2, 3$$

and

$$n_1 + n_2 + n_3 \geq n + 1 - g(n).$$

By definition, $\Sigma g(n_i) \geq \Sigma h(n_i)$. Since $h(r)$ is increasing and concave for $r > 4$, $\Sigma h(n_i)$ is minimized, subject to the preceding constraints on the n_i, when n_1 and n_2 are made as small as possible, and $n_3 = n + 1 - g(n) - (n_1 + n_2)$. Since $-g(n) \geq -(n/(\log n) + 1)$,

$$n_i \geq \frac{n + 2 - 2\left(\frac{n}{\log n} + 1\right)}{3} = \frac{n - \frac{2n}{\log n}}{3}$$

for $i = 1, 2$, and $n_3 \geq n + 1 - [n/(\log n) + 1] - (n_1 + n_2)$. This lower bound on n_3 equals $[n + n/\log n]/3$ when $n_1 = n_2 = (n - 2n/\log n)/3$. It follows that

$$\Sigma h(n_i) \geq 2h\left(\frac{n - 2n/\log n}{3}\right) + h\left(\frac{n + n/\log n}{3}\right).$$

Hence, if the right-hand side of this inequality exceeds $h(n)$, then $\Sigma g(n_i) > n/\log n$ and the proof is complete.

Thus, it remains to show that

$$\frac{2(n - 2n/\log n)/3}{\log\left(\frac{n - 2n/\log n}{3}\right)} + \frac{(n + n/\log n)/3}{\log\left(\frac{n + n/\log n}{3}\right)} > \frac{n}{\log n}$$

for $n \geq 32$. After cancellation and rearrangement, this inequality can be written as

$$3[(\log 3) - 1]\log n + (\log n)\log\left\{\frac{(\log n)^3}{(\log n)^3 - [(\log n)^2 - 2(\log n) - 4]}\right\}$$
$$> 3[(\log 3) - 1]\log 3 + [3(\log 3) + 1]\log(\log n/[(\log n) - 2])$$
$$- \log\left(1 + \frac{1}{\log n}\right)\left\{3\log\left(\frac{\log n}{(\log n) - 2}\right) + [3(\log 3) - 4]\right\}.$$

For $n \geq 32$, the left-hand side exceeds its first term and the right-hand side is less than the sum of its first two terms. Therefore, the inequality holds if

$$\log n > \log 3 + \left\{\frac{3(\log 3) + 1}{3[(\log 3) - 1]}\right\}\log\left[\frac{\log n}{(\log n) - 2}\right].$$

This is true when $n = 32$ and hence for $n > 32$ since its left-hand side increases in n while its right-hand side decreases in n. □

We now develop an upper bound on $\tau(n)$ that for large n is considerably sharper than the bound in Theorem 1. We shall use the following key lemma to prove Theorem 4 and will then consider a construction that leads to a proof of the lemma.

Lemma 5. $\tau(2^k(k+4)) \leq 3(2^k)$ *for* $k = 0, 1, 2, \ldots$, *and the resultant upper bound on* $\tau(n)(\log n)/n$ *for* $n \in \{2^k(k+4): k = 0, 1, \ldots\}$ *converges to 3.*

Theorem 4. *For each* $\delta > 0$, *there is an* $n(\delta)$ *such that* $\tau(n) < (6 + \delta)n/\log_2 n$ *for all* $n > n(\delta)$. *Moreover,*

$$\tau(n) < \frac{7n}{\log_2 n} \quad \text{for all } n \geq 2.$$

Proof of Theorem 4. Let $\alpha_k = 2^k(k + 4)$. If $\alpha_k \leq n \leq \alpha_{k+1}$, then, by Lemma 5,

$$\tau(n) \leq \tau(\alpha_{k+1}) \leq 3(2^{k+1}) = 6(2^k) < 7(2^k)(k + 4)/[k + \log(k + 4)]$$
$$= 7\alpha_k/\log \alpha_k \leq 7n/\log n,$$

so $\tau(n) < 7n/\log n$ for all $n \geq \alpha_0 = 4$. The same bound holds for $n \in \{2, 3\}$.

When 7 in the preceding inequality series is replaced by $6 + \delta$, we get $6(2^k) < (6 + \delta)2^k(k + 4)/[k + \log(k + 4)]$ for sufficiently large k, hence $\tau(n) < (6 + \delta)n/\log n$ for sufficiently large n. □

To approach Lemma 5, we construct a symmetric hierarchical series $P_0, P_1, P_2, \ldots$ of successively larger interval orders such that the largest semiorder in each P_k is obtained (among other ways) both by a chain and by two antichains, much as in Fig. 6.2. The first interval order in the series, P_0, consists of an $[(n/2) + 1]$-point chain and $(n/2) - 1$ other points that bear $\sim$ to everything else. That is, P_0 is one of the p_n as used in the proof of Lemma 1: see Fig. 6.2(a).

Beginning with such a P_0, we construct $P_1, P_2, \ldots$ as follows. Given P_k, its successor P_{k+1} consists of two copies of P_k, one completely to the left of the other (copy 1 $\prec$ copy 2), plus δ_k other points that bear $\sim$ to everything else, such that

$\delta_k + 2(\text{maximum number of points in an antichain of } P_k)$
$= \text{maximum number of points in two antichains of } P_{k+1}$
$= \text{number of points in the maximum chain of } P_{k+1}$
$= 2(\text{number of points in the maximum chain of } P_k).$

Figure 6.5 illustrates this construction when P_0 is the $K_{1,3}^*$ set p_4. The figure shows P_3 as built from two copies of P_2, or four copies of P_1, or eight copies of P_0. It should be apparent from the figure that if the δ_2 points are used for a semiorder in P_3, then the largest such semiorder consists of two antichains containing $8 + 2(4 + 2 + 1 + 1) = 24$ points. On the other hand, if none of the δ_2 points are used for a semiorder in P_3, then the largest such semiorder has twice as many points as the largest semiorder in P_2, namely $2(12) = 24$. The latter maximum can be realized in several ways, one of which is the 24-point chain at the bottom of the figure. Since there are 56 points for P_3, $\tau(56) \leq 24$.

Bearing in mind that $P_0, P_1, P_2, \ldots$ depends on the choice of P_0, let

$\alpha_k = $ number of points in P_k,
$\beta_k = $ number of points in a maximum antichain in P_k,
$\gamma_k = $ number of points in a maximum semiorder in P_k,
$\delta_k = $ number of points added to the two copies of P_k to get P_{k+1}.

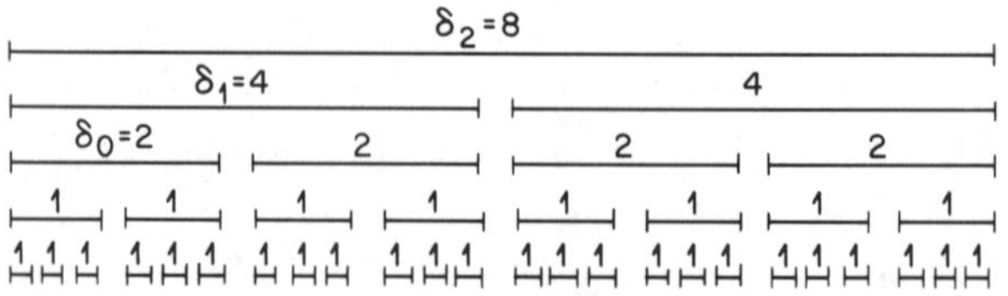

Figure 6.5 P_3 based on $P_0 = K_{1,3}^*$.

General Bounds

The equations in the preceding paragraph give

$$\delta_k + 2\beta_k = \gamma_{k+1} = 2\gamma_k.$$

According to this and the construction,

$$\alpha_{k+1} = 2\alpha_k + \delta_k,$$
$$\beta_{k+1} = \beta_k + \delta_k,$$
$$\gamma_{k+1} = 2\gamma_k = 2\beta_k + \delta_k,$$
$$\delta_{k+1} = 2(\gamma_{k+1} - \beta_{k+1}) = 2\beta_k.$$

Proof of Lemma 5. Let $P_0 = p_4 = K_{1,3}^*$ so that $(\alpha_0, \beta_0, \delta_0, \gamma_0) = (4, 2, 3, 2)$. It then follows from the preceding equation scheme that $\gamma_k = 3(2^k)$ and $\alpha_k = 2^k(k+4)$. Hence $\tau(2^k(k+4)) \le 3(2^k)$. For the convergence part of the lemma, take $n = 2^k(k+4)$. Then $\log n = k + \log(k+4)$, so

$$\frac{\log n}{n}[3(2^k)] = \frac{k + \log(k+4)}{2^k(k+4)}[3(2^k)] \to 3. \qquad \square$$

Our proof of Lemma 5 was based on the simplest P_0 since the successive terms in the recursive scheme preceding the proof are easiest to compute in this case. However, similar conclusions obtain for other P_0 choices. In particular, the bound on $\tau(n)(\log n)/n$ always converges to 3. To see this, we note (proof left to the reader) that, when $P_0 = p_m$, the recursive scheme yields

$$\gamma_k = 2^k \gamma_0$$
$$\beta_k = 2[2^k + (-1)^{k+1}]\gamma_0/3 + (-1)^k \beta_0,$$
$$\delta_k = 2[2^k + (-1)^{k+1}]\gamma_0/3 + (-1)^k \gamma_0,$$

and

$$\alpha_k = 2^k \alpha_0 + [2^k + (-1)^{k+1}]\delta_0/3 + 2[(3k-2)2^{k-1} + (-1)^k]\gamma_0/9$$
$$= 2^k[\alpha_0 + \delta_0/3 + (3k-2)\gamma_0/9] + (-1)^{k+1}(\delta_0/3 - 2\gamma_0/9).$$

With $n = \alpha_k$,

$$\tau(n)(\log n)/n \le \gamma_k(\log \alpha_k)/\alpha_k \to 3.$$

Because of the trade-off and symmetry between chains and double antichains in the construction used for the proof of Lemma 5, it is tempting to conjecture that $\tau(2^k(k+4)) = 3(2^k)$ for $k = 0, 1, 2, \ldots$. Our earlier results show only that equality holds for $k \in \{0, 1\}$: $\tau(4) = 3$ and $\tau(10) = 6$.

It may also be noted as an immediate corollary of Theorem 4 that $\tau(n)/n \to 0$ as n gets large.

6.4 PATTERNS WITH RESTRICTED ANTICHAINS

Theorems 1 and 2 show that τ has a simple repeating pattern on two-integer blocks for small n; that is, $(\tau(3), \tau(4)) = (3, 3)$, $(\tau(5), \tau(6)) = (4, 4), \ldots$, $(\tau(15), \tau(16)) = (9, 9)$. However, since $\tau(17) \neq 10$, the pattern does not continue beyond $(\tau(15), \tau(16))$. In contrast to this, we shall now prove that when all interval orders under consideration are restricted to have no antichain with more than j points, for small j, then there is a general repeating pattern for the $\tau_j(n)$. Recall that $\tau_j(n)$ is the largest k such that every n-point interval order with no antichain of more than j points has a k-point semiorder embedded within it.

The situation for $j = 1$ is trivial since $\tau_1(n) = n$. The following theorems show what happens for $j = 2$ and $j = 3$. The question of whether there are indefinite repeating patterns for $j \geq 4$ is open.

Theorem 5. *Let $(c_1, c_2, c_3, c_4) = (1, 2, 3, 3)$. Then, for all $n \in \{0, 1, \ldots\}$ and all $i \in \{1, 2, 3, 4\}$,*

$$\tau_2(4n + i) = 3n + c_i.$$

Theorem 6. *Let $(c_1, c_2, c_3, c_4, c_5) = (1, 2, 3, 3, 4)$. Then, for all $n \in \{0, 1, \ldots\}$ and all $i \in \{1, \ldots, 5\}$,*

$$\tau_3(5n + i) = 3n + c_i.$$

The pattern of τ_2 values is $1, 2, 3, 3; 4, 5, 6, 6; 7, 8, 9, 9; \ldots$, and the pattern of τ_3 values is $1, 2, 3, 3, 4; 4, 5, 6, 6, 7; 7, 8, 9, 9, 10; \ldots$. The proof of Theorem 5 is quite short and will be given first. The proof of Theorem 6, which involves an interesting induction argument applied to each of four disjoint subsequences of the positive integers, will conclude the chapter.

Proof of Theorem 5. Figure 6.6 pictures an interval order on $4n + i$ points that has no three-point antichain. Since each of the four-interval blocks

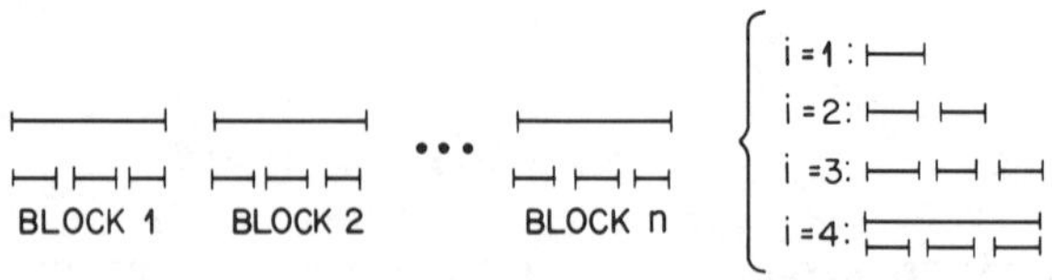

Figure 6.6 Interval order with no three-point antichain.

contributes exactly three points to a maximum semiorder,

$$\tau_2(4n + i) \le 3n + c_i, \qquad [n = 0, 1, \ldots; i = 1, 2, 3, 4]$$

where $(c_1, c_2, c_3, c_4) = (1, 2, 3, 3)$.

Equality obviously holds in the preceding inequality if $n = 0$. Assuming that equality holds through $n \ge 0$, we show that it holds at $n + 1$. Given an interval order on $4(n + 1) + i$ points $(1 \le i \le 4)$ that has no three-point antichain, assume for definiteness that a representation F has $F_1^- \le F_2^- \le \cdots \le F_{4(n+1)+i}^-$, where the notation is like that used in the proofs of Lemmas 2 through 4. Let $h = |\{i: F_i^+ < F_5^-\}|$. Because there is no three-point antichain, $h \in \{3, 4\}$. Therefore at least three intervals are strictly to the left of intervals F_5 through $F_{4(n+1)+i}$, and it follows from the induction hypothesis that

$$\tau_2(4(n + 1) + i) \ge \tau_2(3) + \tau_2(4(n + 1) + i - 4)$$
$$= 3 + \tau_2(4n + i) = 3 + 3n + c_i = 3(n + 1) + c_i. \quad \square$$

Proof of Theorem 6. It is easily seen that $\tau_3(i) = c_i$ for $1 \le i \le 5$, where $(c_1, c_2, c_3, c_4, c_5) = (1, 2, 3, 3, 4)$. We therefore assume the conclusion of Theorem 6 for $n = 0$. With c_i as defined here, we first establish the desired lower bound on τ_3 and then show that this bound is tight.

Lemma 6. $\tau_3(5n + i) \ge 3n + c_i$.

Proof. Assume that Lemma 6 holds through $n \ge 0$. To verify it at $n + 1$, order the intervals of a representation of an interval order on $5(n + 1) + i$ points $(1 \le i \le 5)$ with no four-point antichain as $F_1^- \le \cdots \le F_{5(n+1)+i}^-$. Because there is no four-point antichain, at least $m - 2$ of the intervals for points 1 through m completely precede the intervals for points $m + 1$ through $5(n + 1) + i$, and therefore

$$\tau_3(5(n + 1) + i) \ge \tau_3(m - 2) + \tau_3(5(n + 1) + i - m).$$

We shall use this fact in the cases for $i \in \{2, 3, 5\}$ in what follows.

$i = 1$. Since τ_3 is nondecreasing, $\tau_3(5(n + 1) + 1) \ge \tau_3(5n + 5) \ge 3n + c_5 = 3(n + 1) + c_1$ since $c_5 = 4$ and $c_1 = 1$.

$i = 2$. Set $m = 4$. Then $\tau_3(5(n + 1) + 2) \ge \tau_3(2) + \tau_3(5n + 3) \ge 2 + (3n + c_3) = 3(n + 1) + c_2$ since $c_3 = 3$ and $c_2 = 2$.

$i = 3$. Set $m = 5$. Then $\tau_3(5(n + 1) + 3) \ge \tau_3(3) + \tau_3(5n + 3) \ge 3 + (3n + c_3) = 3(n + 1) + c_3$.

$i = 4$. $\tau_3(5(n + 1) + 4) \ge \tau_3(5(n + 1) + 3) \ge 3(n + 1) + c_4$ since $c_4 = c_3$.

$i = 5$. Set $m = 5$. Then $\tau_3(5(n + 1) + 5) \ge \tau_3(3) + \tau_3(5n + 5) \ge 3 + (3n + c_5) = 3(n + 1) + c_5$. $\square$

Our first step in showing that the bound in Lemma 6 is tight involves the following construction, which is presumed for Lemma 7.

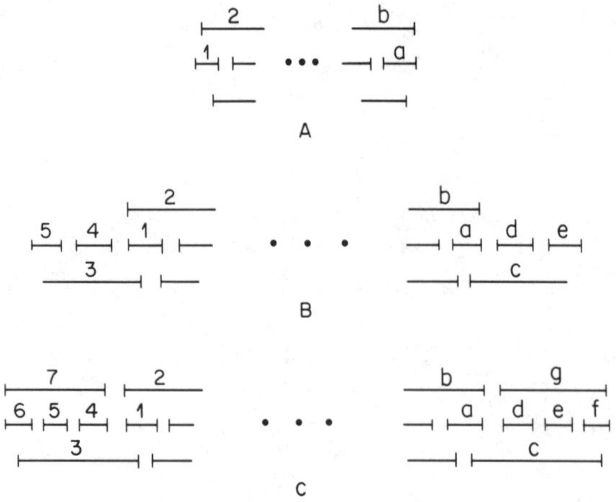

Figure 6.7 Nested hierarchy of interval orders.

Let A, B, and C be successively inclusive interval orders on k, $k + 6$, and $k + 10$ points, respectively: B is formed from A by adding six points to A without changing any $\prec$ or $\sim$ ($= sc(\prec)$) relationships within A, and C is formed from B by adding four points to B so that the restriction of C to the point set of B is B. Figure 6.7 illustrates the additions. It is assumed that $k \geq 4$. The interval representation for A involves a chain on points 1 through a ($a \neq 1$), plus intervals for points 2 and b that, respectively, intersect 1 and a (we require $\{2, b\} \cap \{1, a\} = \emptyset$, but allow $2 = b$), plus other intervals. All of these others begin after 1 and 2 begin and end before a and b end.

To get B from A, add points 3, 4, and 5 on the left of A so that $5 \prec 4 \prec \{1, 2\}$ with 3's interval intersecting those for 5, 4, 2, and 1 (but no others). Also add points c, d, and e at the right end of A in a symmetric manner. To get C from B, add 6 and 7 at the left of B and f and g at the right of B as shown in the figure, with 3 and c extended if necessary to intersect 6 and 7, or f and g.

Let $\lambda(A)$ be the number of points in a maximum semiorder in A, and define $\lambda(B)$ and $\lambda(C)$ similarly.

Lemma 7. *Suppose every semiorder in A that has $\lambda(A)$ points contains either point 1 or 2, and either point a or b, and every semiorder in A that has $\lambda(A) - 1$ points contains at least one point in $\{1, 2, a, b\}$. Then*

$$\lambda(B) = \lambda(A) + 4, \quad \lambda(C) = \lambda(B) + 2 = \lambda(A) + 6;$$

every semiorder in C that has $\lambda(C)$ points contains either point 6 or 7, and either point f or g; and every semiorder in C that has $\lambda(C) - 1$ points contains at least one point in $\{6, 7, f, g\}$.

Remark. If A includes no antichain of four or more points, then neither does B or C. In the induction proof which follows the proof of Lemma 7, C

takes the place of A in the "next step" with points 6, 7, f, and g serving the roles formerly played by 1, 2, a, and b.

Proof of Lemma 7. Given A with $k \geq 4$ points, the hypotheses of Lemma 7 are assumed throughout its proof. For definiteness, assume also that each semiorder within A that has $\lambda(A) - 1$ points in the ensuing treatment contains a point in $\{1,2\}$.

Given a $\lambda(A)$-point semiorder in A, we can add two but not three points from $\{3,4,5\}$, say 4 and 5 for example, and add two but not three points from $\{c, d, e\}$ to that semiorder to obtain a semiorder within B that has $\lambda(A) + 4$ points. A semiorder on $\lambda(A) - 1$ points within A can be augmented by at most two points from $\{3,4,5\}$ and perhaps all three of c, d, and e to yield a semiorder in B of no more than $\lambda(A) - 1 + 5 = \lambda(A) + 4$ points. It follows that $\lambda(B) = \lambda(A) + 4$.

Consider a $\lambda(B)$-point semiorder in B:

(i) If its restriction on A has $\lambda(A)$ points, then it contains 1 or 2, two in $\{3,4,5\}$, a or b, and two in $\{c,d,e\}$.

(ii) If its restriction on A has $\lambda(A) - 1$ points, then it contains 1 or 2, two in $\{3,4,5\}$, none in $\{a,b\}$, and all three in $\{c,d,e\}$.

(iii) If its restriction on A has $\lambda(A) - 2$ points, then the semiorder from B contains 3, 4, 5, c, d, and e.

Regardless of which of (i) through (iii) holds, it is easily seen that the $\lambda(B)$-point semiorder in B can be augmented by at most one point in $\{6,7\}$ and at most one point in $\{f, g\}$ to yield a larger semiorder in C. Moreover, we can always add at least one point from each of $\{6,7\}$ and $\{f, g\}$ to get an expanded semiorder in C. Therefore

(iv) If the restriction on B of a semiorder in C has $\lambda(B)$ points, then the semiorder in C has either $\lambda(B) + 2$ points (in which case it contains 6 or 7, and f or g) or $\lambda(B) + 1$ points (in which case it contains one point from $\{6,7,f,g\}$) or $\lambda(B)$ points.

Consider next a semiorder in B of $\lambda(B) - 1 = \lambda(A) + 3$ points. Regardless of whether its restriction on A has $\lambda(A)$, $\lambda(A) - 1$, $\lambda(A) - 2$, or $\lambda(A) - 3$ points, it is easily seen that at most three points from $\{6,7,f,g\}$ can be added to it to yield a larger semiorder in C. Therefore

(v) If the restriction on B of a semiorder in C has $\lambda(B) - 1$ points, then the semiorder in C has either $\lambda(B) + 2$ points (including 6 or 7, and f or g) or $\lambda(B) + 1$ points (including two in $\{6,7,f,g\}$) or fewer than $\lambda(B) + 1$ points.

In addition to (iv) and (v), it is obvious that if the restriction on B of a semiorder in C has $\lambda(B) - 2 = \lambda(A) + 2$ points, then the semiorder in C has $\lambda(B) + 2$ points (including all in $\{6,7,f,g\}$) or $\lambda(B) + 1$ points (including three in $\{6,7,f,g\}$) or fewer than $\lambda(B) + 1$ points. Moreover, if the restriction on B of a semiorder in C has $\lambda(B) - 3$ points, then the semiorder in C can have at most $\lambda(B) + 1$ points (which include all of 6, 7, f, and g).

It follows that $\lambda(C) = \lambda(B) + 2$ and that the assertions in the final part of Lemma 7 are true. $\square$

We now complete the proof of Theorem 6. This will be done by applying Lemma 7 to the following subsequences of integers whose successive terms have alternating increases of 6 and 4 in adherence with the pattern developed following the proof of Lemma 6:

S1. 4, 10, 14, 20, 24, ...
S2. 6, 12, 16, 22, 26, ...
S3. 9, 15, 19, 25, 29, ...
S4. 11, 17, 21, 27, 31, ...

An interval order that serves within Lemma 7 for the first term in each subsequence is shown in Fig. 6.8. Each A has no four-point antichain and is easily seen to satisfy the hypotheses of the lemma. The maximum semiorders in the four cases have, respectively, 3, 4, 6, and 7 points, which in view of Lemma 6 verify

$$\tau_3(4) = 3, \qquad \tau_3(6) = 4, \qquad \tau_3(9) = 6, \qquad \tau_3(11) = 7.$$

To check the A for S4 against the hypotheses of Lemma 7, note first that if neither a nor b is used in a semiorder, then the semiorder has at most six points. Second, if none of 1, 2, a, and b is used in a semiorder, then it has at most five points.

With $\tau_3(4) = 3$ for the first term in S1, successive applications of Lemma 7 on the terms in S1 give $\tau_3(10) \leq 3 + 4 = 7$ and $\tau_3(14) \leq 3 + 6 = 9$, then $\tau_3(20) \leq 9 + 4$ and $\tau_3(24) \leq 9 + 6$, and so forth.

S1. $\tau_3(4 + 10n) \leq 3 + 6n$ for $n = 0, 1, ...$
     $\tau_3(10 + 10n) \leq 7 + 6n$ for $n = 0, 1, ...$.

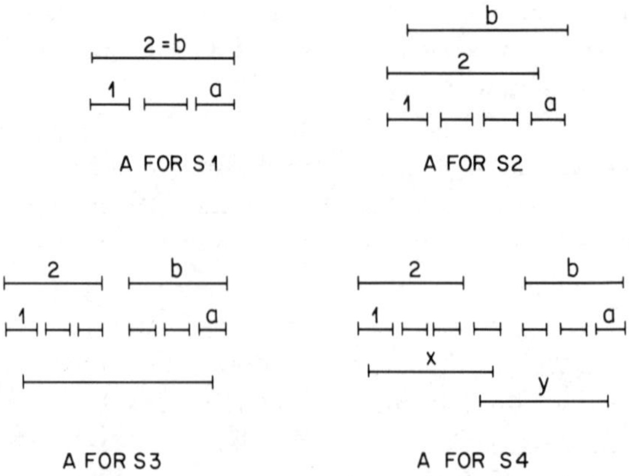

Figure 6.8 Interval orders with no four-point antichains.

Similar applications of Lemma 7 to the other Sk give

S2. $\tau_3(6 + 10n) \le 4 + 6n$ and $\tau_3(12 + 10n) \le 8 + 6n$,

S3. $\tau_3(9 + 10n) \le 6 + 6n$ and $\tau_3(15 + 10n) \le 10 + 6n$,

S4. $\tau_3(11 + 10n) \le 7 + 6n$ and $\tau_3(17 + 10n) \le 11 + 6n$

for $n = 0, 1, 2, \ldots$. The first parts of S2 and S4 show that

$$\tau_3(5n + 1) \le 3n + 1 \quad \text{for} \quad n = 0, 1, \ldots .$$

The second parts of S2 and S4 give (in view of $\tau_3(2) = 2$ and the easily verified $\tau_3(7) \le 5$)

$$\tau_3(5n + 2) \le 3n + 2.$$

Monotonicity of τ_3 and the first parts of S1 and S3 yield

$$\tau_3(5n + 3) \le \tau_3(5n + 4) \le 3n + 3,$$

and the second parts of S1 and S3 (plus $\tau_3(5) = 4$) give

$$\tau_3(5n + 5) \le 3n + 4.$$

Therefore $\tau_3(5n + i) \le 3n + c_i$ for $n = 0, 1, \ldots$ and $i = 1, 2, 3, 4, 5$. The conclusion of Theorem 6 follows immediately from this and Lemma 6. □

7

Real Representations

This chapter continues the discussion of Chapter 2 on representations of interval orders and semiorders by families of real intervals. It begins with a constructive proof of the Scott-Suppes semiorder theorem, then shows how solution theory for finite sets of linear inequalities can be used to prove the existence of representations for various types of interval orders and semiorders. The latter part of the chapter discusses real representations for interval orders of arbitrary cardinality, based in part on Cantor's theorem for the existence of an isomorphism between a linearly ordered set and a subset of $(\mathbb{R}, <)$.

7.1 THE SCOTT-SUPPES THEOREM

It is assumed throughout the chapter that $\sim = sc(\prec)$ and that $x \approx y \Leftrightarrow \{z: z \sim x\} = \{z: z \sim y\}$. The present section and the next two sections work with finite, nonempty X.

Our immediate purpose is to give a constructive proof of

Theorem 2.9. *If $(X, \prec)$ is a finite semiorder, then there is an $f: X \to \mathbb{R}$ such that, for all $x, y \in X$,*

$$x \prec y \Leftrightarrow f(x) + 1 < f(y).$$

An existence proof of this theorem is noted in the next section. Other constructive proofs appear in Rabinovitch (1977) and Roberts (1979) as well as in Scott and Suppes (1958). The following proof is close to the original.

Proof. It is easily seen that it suffices to prove the theorem when $x \approx y \Rightarrow x = y$ and $(X, \sim)$ is connected, so assume these for semiorder $(\{x_1, \ldots, x_n\}, \prec)$. Since $<^-$ as defined in Section 5.4 is a linear order on $\{x_1, \ldots, x_n\}$ ($<^- = <^+ = [\prec^- \cup (\sim^- \cap \prec^+)] = [\prec^+ \cup (\sim^+ \cap \prec^-)]$), assume with no loss in generality that $x_1 <^- x_2 <^- \cdots <^- x_n$. Let

$$r_i = \max\{j: x_j \sim x_i\} \quad \text{for } i = 1, \ldots, n.$$

It is easily checked that $r_i > i$ for all $i < n$, and that $1 < r_1 \leq r_2 \leq \cdots \leq r_{n-1} = r_n = n$.

We now define $f(i) = f(x_i)$ to satisfy the conclusion of Theorem 2.9. For nontriviality, take $n \geq 3$. Set

$f(1) = 0$, $f(r_1) = 1$ and, if $r_1 \geq 3$, assign increasing values to $f(2), \ldots, f(r_1 - 1)$ that are strictly between 0 and 1.

Since $x_i \sim x_j$ whenever $i, j \leq r_1$, the assignment thus far satisfies the theorem.

Suppose $f(1)$ through $f(r_i)$ have been assigned so that the representation holds on $\{x_1, \ldots, x_{r_i}\}$ with $i \geq 1$ and $r_i < n$. Let $k = \min\{j: j > i$ and $r_j > r_i\}$. Set

$f(r_k) = f(k) + 1$ and, if $r_k > r_i + 1$, assign increasing values to $f(r_i + 1), \ldots, f(r_k - 1)$ that are strictly between $f(k - 1) + 1$ and $f(r_k)$.

By the definition of k, $x_a \prec x_b$ when $a \leq k - 1$ and $r_i < b \leq r_k$, in which case $f(a) \leq f(k - 1) < f(b) - 1$, so $f(a) + 1 < f(b)$. Moreover, $x_a \sim x_b$ when $k \leq a < b \leq r_k$, and in this case $f(r_k) - 1 = f(k) \leq f(a) < f(b) \leq f(r_k)$, so $f(b) - f(a) \leq 1$. Therefore the conclusion of the theorem holds on $\{x_1, \ldots, x_{r_k}\}$.

Continuation through the first $r_j = n$ completes the construction. □

7.2 LINEAR SOLUTION THEORY

This section discusses an important theorem for the existence of a solution to a finite set of linear inequalities, then shows how it implies that there exists a representation (f, ρ) of a finite interval order for a given positive length function ρ if, and only if, ρ satisfies a certain set of inequalities. An alternative proof of Theorem 2.9 follows easily from the latter result, as noted at the end of the section. The linear solution-existence theorem, Theorem 1, will also be used in the next section and in the next chapter.

We shall need a few definitions. Elements in a finite-dimensional Euclidean space $\mathbb{R}^n$ will be referred to as *vectors*. Components of vectors will be denoted by subscripts: $x^k = (x_1^k, \ldots, x_n^k)$. A vector is *rational* if all its components are rational numbers, and is *integral* if all its components are integers. For real λ and vectors $x, y \in \mathbb{R}^n$, $\lambda x = (\lambda x_1, \ldots, \lambda x_n)$, $x + y = (x_1 + y_1, \ldots, x_n + y_n)$, and

$$x \cdot y = \sum_{i=1}^{n} x_i y_i.$$

Other notions will be introduced when needed.

Many mathematicians have been involved in the development of linear solution theory. Important works include Farkas (1902), Klee (1955), Kuhn (1956), and the articles by Ky Fan, A. J. Goldman, and A. W. Tucker in Kuhn and Tucker (1956). The following theorem, or a closely related theorem, is sometimes referred to as Farkas's lemma or as the theorem of the alternative. It is intimately related to standard separating-hyperplane theorems, or linear separation theorems. We assume that $n \in \{1, 2, \ldots\}$.

Theorem 1. *Suppose $x^1, \ldots, x^N \in \mathbb{R}^n$ and $1 \leq K \leq N$. Then exactly one of* **A** *and* **B** *is true*:

A. *There is a $w \in \mathbb{R}^n$ such that*

$$w \cdot x^k > 0 \quad \text{for } k = 1, \ldots, K,$$
$$w \cdot x^k \geq 0 \quad \text{for } k = K+1, \ldots, N;$$

B. *There are nonnegative real r_k for $k = 1, \ldots, N$ with $r_1 + \cdots + r_K > 0$ such that*

$$\sum_{k=1}^{N} r_k x_i^k = 0 \quad \text{for } i = 1, \ldots, n.$$

If, in addition, the x^k are all rational, then there is an integral w that satisfies **A** *when* **A** *holds, and there are integers $r_1, \ldots, r_n$ that satisfy* **B** *when* **B** *holds.*

Proof (outline). Given the hypotheses, let $y^k = -x^k$ for $k > K$, let C_1 be the convex hull of $\{x^1, \ldots, x^K\}$, and let C_2 be the convex cone with origin $\mathbf{0} \in \mathbb{R}^n$ generated by $\{y^{K+1}, \ldots, y^N\}$:

$$C_1 = \left\{ \sum_{k=1}^{K} \lambda_k x^k : \lambda_k \geq 0 \text{ for } k \leq K, \sum \lambda_k = 1 \right\},$$

$$C_2 = \mathbf{0} \cup \left\{ \sum_{k=K+1}^{N} \mu_k y^k : \mu_k \geq 0 \text{ for } k > K, \sum \mu_k > 0 \ (N > K) \right\}.$$

Since both **A** and **B** cannot hold, else $0 = w \cdot (\Sigma r_k x^k) = \Sigma r_k (w \cdot x^k) > 0$, the proof of the main part of the theorem involves showing that $C_1 \cap C_2 = \emptyset \Rightarrow$ **A** and $C_1 \cap C_2 \neq \emptyset \Rightarrow$ **B**.

Suppose $C_1 \cap C_2 = \emptyset$. Then, since the x^k sets used to generate the closed convex sets C_1 and C_2 are finite, linear separation theory (see, for example, Klee, 1955; Kuhn and Tucker, 1956; Rockafellar, 1970) implies that there is a minimum positive distance between C_1 and C_2 achieved by points c_1 and c_2 on the boundaries of C_1 and C_2, respectively. It is not hard to show that $w = c_1 - c_2$ satisfies **A**.

On the other hand, if $z \in C_1 \cap C_2$ with $z = \Sigma \lambda_k x^k = \Sigma \mu_k y^k$ (or $z = \mathbf{0}$), then $\Sigma_{k \leq K} \lambda_k x^k + \Sigma_{k > K} \mu_k(-y^k) = \mathbf{0}$, so the λ_k and μ_k suffice as the r_k for **B**.

Suppose all x^k are rational. If **B** holds, let r be a solution with the fewest positive r_k, say r_1 through r_J, and consider a maximum set of linearly independent equations in $\{\Sigma_{k \leq J} r_k x_i^k = 0;\ i = 1, \ldots, N\}$. If $J = 1$, any $r_1 > 0$ will do since then $x^1 = \mathbf{0}$, so suppose $J > 2$. Then standard solution theory for systems of homogeneous linear equations shows that the positive r_k, if not already rational, can be perturbed slightly to give a positive rational solution. If **A** holds, a similar perturbation on the w_i can be made for all equations $w \cdot x^k = 0$ (if any) without altering the equalities and without changing the strict inequalities for the $w \cdot x^k > 0$. □

Our applications of Theorem 1 will use the final part of the theorem for rational x^k and will follow the common procedure of proving that **A** holds by showing that **B** is impossible in the case at hand. In particular, the supposition that **B** holds yields nonnegative integers r_k, at least one of which is positive, which are then used to derive a contradiction. We shall also be primarily concerned with the strict inequality part of **A**.

The first application involves an extension of the relation $\subset_0$ on X, defined in Section 5.1 for the depth of a poset, to a relation on nonempty subsets of X. Recall that when $(X, \prec)$ is an interval order and $x \subset_0 y$, then $\rho(x) < \rho(y)$ for every representation (f, ρ) of $(X, \prec)$. The converse is true also. For the extension, we define $\subset_0$ on pairs of nonempty disjoint subsets A and B in X by

$$A \subset_0 B \quad \text{if} \quad \sum_{a \in A} \rho(a) < \sum_{b \in B} \rho(b)$$

for every representation (f, ρ) of a finite interval order $(X, \prec)$ for which $\rho > 0$. Hence $A \subset_0 B$ if and only if the sum of the lengths used for intervals for A *must* be less than the sum of the lengths used for intervals for B.

Figure 7.1 pictures two examples of $A \subset_0 B$. The top case requires $\rho(a_1) + \rho(a_2) < \Sigma \rho(b_i)$ since the disjoint a_i intervals lie completely within the span of the overlapping b_i intervals. The bottom case requires $\rho(a_1) + \rho(a_2) < \rho(b_1) + \rho(b_2)$ since a_1 and a_2 are properly included within b_2 and the intersection of a_1 and a_2 is properly included within b_1.

We shall refer to the set of inequalities

$$\left\{ \sum_A \rho(a) < \sum_B \rho(b) : A \subset_0 B\ (A \neq \varnothing, B \neq \varnothing, A \cap B = \varnothing) \right\}$$

for a finite interval order $(X, \prec)$ as its *ρ-set*. By definition, if (f, ρ) is a representation of $(X, \prec)$, then ρ satisfies the ρ-set of $(X, \prec)$, that is, all inequalities in the ρ-set hold for the given length function. The interesting fact is that the converse is true.

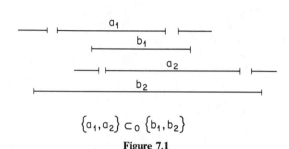

Figure 7.1

Theorem 2. *Suppose* $(X, \prec)$ *is a finite interval order. Then, for every positive length function ρ on X, there is a location function f on X such that (f, ρ) is a representation of $(X, \prec)$ if, and only if, ρ satisfies the ρ-set of $(X, \prec)$.*

Proof. Let $\rho > 0$ on X be given for a finite interval order $(X, \prec)$. We are to show that if ρ satisfies the ρ-set then there is an f such that (f, ρ) is a representation of $(X, \prec)$. This will be done by proving that if there is no f for which (f, ρ) is a representation, then ρ violates some inequality in the ρ-set.

For convenience, we shall work with the dual $\succ$ of $\prec$. Let $X = \{y_1, y_2, \ldots, y_n\}$ with $|X| = n$. Given ρ, the inequalities that must hold for (f, ρ) to be a representation are:

$$f(y_i) - f(y_j) - \rho(y_j) > 0 \quad \text{whenever } y_i \succ y_j;$$
$$f(y_i) - f(y_j) + \rho(y_i) \geq 0 \quad \text{whenever } y_i \sim y_j \text{ and } i \neq j.$$

Define a coefficient vector $a(i, j) \in \mathbb{R}^{n+1}$ for each of these inequalities by:

if $y_i \succ y_j$, $a(i, j)$ has 1 in position i, -1 in position j, $-\rho(y_j)$ in position $n + 1$, and 0's elsewhere;

if $y_i \sim y_j$, $a(i, j)$ has 1 in position i, -1 in position j, $\rho(y_i)$ in position $n + 1$, and 0's elsewhere.

Also let $w' = (f(y_1), f(y_2), \ldots, f(y_n), 1)$ in $\mathbb{R}^{n+1}$. The preceding system of inequalities is therefore

$$w' \cdot a(i, j) > 0 \quad \text{whenever } y_i \succ y_j;$$
$$w' \cdot a(i, j) \geq 0 \quad \text{whenever } y_i \sim y_j \text{ and } i \neq j.$$

Linear Solution Theory

Because ρ can be multiplied by an arbitrary positive constant without affecting the existence of an f solution, it follows that there is an f for which (f, ρ) is a representation if and only if there is a $w \in \mathbb{R}^{n+1}$ that satisfies the following system:

$$w \cdot (0, \ldots, 0, 1) > 0 \quad [\text{so } w_{n+1} > 0]$$

$$w \cdot a(i, j) > 0 \quad \text{whenever } y_i \succ y_j,$$

$$w \cdot a(i, j) \geq 0 \quad \text{whenever } y_i \sim y_j \text{ and } i \neq j.$$

Suppose there are K $y_i \succ y_j$ inequalities and $N - K$ ($y_i \sim y_j, i \neq j$) inequalities. Let $a^0 = (0, \ldots, 0, 1)$ and let a_k for $k \geq 1$ correspond to the $a(i, j)$ in a list of the K followed by a list of the $N - K$. Then the system is

$$w \cdot a^0 > 0$$

$$w \cdot a^k > 0 \quad \text{for } k = 1, \ldots, K,$$

$$w \cdot a^k \geq 0 \quad \text{for } k = K+1, \ldots, N.$$

Suppose there is no w solution for this system, that is, no (f, ρ) representation for the given ρ. Then clearly $K \geq 1$ (else $\succ$ is empty) and $N > K$ (else $\succ$ is a chain). Since **A** of Theorem 1 is supposed to be false, it follows that **B** holds: there are nonnegative $r_0, r_1, \ldots, r_N$ with at least one of r_0 through r_K positive such that

$$\sum_{k=1}^{N} r_k a_i^k = 0 \quad \text{for } i = 1, \ldots, n,$$

and

$$r_0 + \sum_{k=1}^{N} r_k a_{n+1}^k = 0.$$

The final equation refers to the final column of the $[a^k]$ matrix, which begins with 1 and has either a $-\rho(y_j)$ or a $+\rho(y_i)$ in every other row. The reason that r_0 does not appear in its predecessors is that $a_i^0 = 0$ for all $i \leq n$. If $r_0 > 0$ then at least one other r_k must be positive according to the final equation. Moreover, because $a_{n+1}^k > 0$ for a $y_i \sim y_j$ case, and $a_{n+1}^k < 0$ for a $y_i \succ y_j$ case, some r_k for $k \in \{1, \ldots, K\}$ must be positive. This in turn leads, by the earlier equations, to the fact that some r_k for $k > K$ must also be positive.

Assume until later that ρ is rational. Then, by the final part of Theorem 1, we can presume that all r_k are integers. We use all $r_k \geq 2$ to replicate the corresponding $y_i \succ y_j$ or $y_i \sim y_j$ r_k times in conjunction with single instances for the $r_k = 1$ to conclude that there is a matrix

$$\begin{bmatrix} a_1 & a_2 & \cdots & a_I & b_1 & b_2 & \cdots & b_J \\ c_1 & c_2 & \cdots & c_I & d_1 & d_2 & \cdots & d_J \end{bmatrix}$$

of points in X with the following five properties:

1. $I \geq 1$, $J \geq 1$.
2. $a_i \succ c_i$ for $i = 1, \ldots, I$.
3. $b_j \sim d_j$ for $j = 1, \ldots, J$.
4. Each $y \in X$ appears the same number of times in each row; that is, $(c_1, \ldots, d_J)$ is a permutation of $(a_1, \ldots, b_J)$.
5. $\Sigma_i \rho(c_i) \geq \Sigma_j \rho(b_j)$.

Properties 1 through 3 follow directly from the observations at the end of the preceding paragraph and the successive listing of the $\succ$ and $\sim$ statements for the positive r_k for $k \geq 1$. In particular, if $r_k \geq 1$ corresponds to $y_i \succ y_j$, then (y_i, y_j) appears as an (a, c) pair r_k times, and $I = \Sigma\{r_k: 1 \leq k \leq K\}$; if $r_k \geq 1$ corresponds to $y_i \sim y_j$, then (y_i, y_j) appears as a (b, d) pair r_k times, and $J = \Sigma\{r_k: k > K\}$. Property 4 is an immediate consequence of the equations $\Sigma r_k a_i^k = 0$ for $i = 1, \ldots, n$, which say that, for each y_i, the number of times y_i is an a or b ($a_i^k = 1$) equals the number of times it is a c or d ($a_i^k = -1$). Finally, Property 5 arises from $r_0 + \Sigma r_k a_{n+1}^k = 0$. Since $r_0 \geq 0$, $\Sigma r_k a_{n+1}^k \leq 0$ with a_{n+1}^k either a $-\rho(c)$ or a $+\rho(b)$.

According to Property 4, the columns of the foregoing matrix can be rearranged into a number of "cyclic blocks," each of which has the form

$$\begin{array}{|ccccc|} t_1 & t_2 & \cdots & t_{H-1} & t_H \\ t_2 & t_3 & \cdots & t_H & t_1 \end{array}$$

with $t_i \neq t_j$ whenever $i \neq j$. Each column of a block inherits either $t_h \succ t_{h+1}$ or $t_h \sim t_{h+1}$ ($t_{H+1} = t_1$) by Properties 2 and 3.

According to Property 5, at least one cyclic block has $\Sigma\{\rho(t_{h+1}): t_h \succ t_{h+1}\} \geq \Sigma\{\rho(t_h): t_h \sim t_{h+1}\}$. For definiteness, let this be the block pictured in the preceding paragraph. We note first that some $\rho(t_h)$ on the right-hand side of the inequality does not appear on the left-hand side. Otherwise, every instance of $t_h \sim t_{h+1}$ (which must occur since $\succ$ is a partial order—else we get $t_1 \succ t_1$) must be preceded by $t_{h-1} \succ t_h$, and contraction by transitivity of adjacent $\succ$ pairs yields a string such as

$$t'_1 \succ t'_2 \sim t'_3 \succ t'_4 \sim t'_5 \succ t'_6 \cdots \succ t'_k \sim t'_1.$$

But, since $\succ$ is assumed to be an interval order, such a string under (2.1) of Section 2.1 gives $t'_1 \succ t'_4$, then $t'_1 \succ t'_6, \ldots$, and finally $t'_1 \succ t'_k \sim t'_1$, for a contradiction. Let B denote the t_h that appear as arguments of ρ only on the right-hand side of $\Sigma\{\rho(t_{h+1}): t_h \succ t_{h+1}\} \geq \Sigma\{\rho(t_h): t_h \sim t_{h+1}\}$, and let A denote the t_{h+1} that appear as arguments of ρ only on the left-hand side, so that

$$\sum_{a \in A} \rho(a) \geq \sum_{b \in B} \rho(b)$$

with $B \neq \emptyset$, as just proved, $A \neq \emptyset$ (else $0 \geq \rho(b)$, contrary to $\rho(b) > 0$), and $A \cap B = \emptyset$ by construction.

We claim that the inequality just derived proves that ρ violates some inequality in the ρ-set of $(X, \prec)$. Indeed, consider the cyclic block that gives rise to $\Sigma_A \rho(a) \geq \Sigma_B \rho(b)$. When we write the equations in (f, ρ) that must hold if (f, ρ) is a representation, that is, $f(t_h) > f(t_{h+1}) + \rho(t_{h+1})$ for $t_h \succ t_{h+1}$, and $f(t_h) + \rho(t_h) \geq f(t_{h+1})$ for $t_h \sim t_{h+1}$, then add these over h and cancel identical f and ρ terms on the two sides, we are left with

$$\sum_{b \in B} \rho(b) > \sum_{a \in A} \rho(a),$$

which must therefore be an inequality in the ρ-set. Thus, our supposition that there is no f for a representation when ρ is rational, implies that ρ violates some inequality in the ρ-set.

This completes the proof for rational ρ. The proof for ρ that have irrational values then follows by arguments of standard analysis. An outline of these arguments will conclude the proof.

With $X = \{y_1, \ldots, y_n\}$, view each $\rho > 0$ as a vector $(\rho(y_1), \ldots, \rho(y_n))$ in $\mathbb{R}^n$. The set of all admissible $\rho' > 0$, that is, those for which there is an (f, ρ') representation, is an open convex cone in $\mathbb{R}^n$. Openness follows from the fact that a given representation (f, ρ') can be transformed into another representation (f', ρ') that gives strict inequalities for all $\succ$ and $\sim$ ($y_i \neq y_j$) statements by taking $f'(y) = f(y) + g(f(y))$ where g is a decreasing function bounded between 0 and some small $\delta > 0$. Then (f', ρ'') will also be a representation for every ρ'' within some small positive distance of ρ', and therefore the set of admissible ρ' is open.

Given any *inadmissible* $\rho > 0$, it follows that there is a hyperplane in $\mathbb{R}^n$ containing ρ such that all $\rho' > 0$ in one of the half spaces determined by the hyperplane are inadmissible. This half space clearly contains a sequence of rational $\rho' > 0$ that converges to ρ. Since every ρ' in this sequence violates one of the inequalities in the finite ρ-set, there must be an inequality in the ρ-set that is violated infinitely often and, by convergence, this inequality will be violated by ρ. □

Existence proofs of the Scott-Suppes semiorder theorem, such as that in Scott (1964), are simpler than the preceding proof because of the additional structure of semiorders. However, with Theorem 2 at hand, we can use it to obtain Theorem 2.9 as an easy corollary. An additional observation for semiorders is needed.

Lemma 1. *Suppose $A \subset_0 B$ for a finite semiorder $(X, \succ)$. Then $|A| < |B|$.*

Proof. Given $A \subset_0 B$ for a finite semiorder $(X, \prec)$, let F be a real representation of the semiorder that satisfies the conditions in Theorem 2.7.

Let

$$A(r) = |\{a \in A : r \in F(a)\}|, \qquad B(r) = |\{b \in B : r \in F(b)\}|.$$

Then $A(r) \leq B(r)$ for every real r, else insertion of an arbitrarily long interval at a point r with $A(r) > B(r)$ gives $\Sigma\rho(a) > \Sigma\rho(b)$; and $A(r) < B(r)$ for some r, else $\Sigma\rho(a) \geq \Sigma\rho(b)$. It follows from these facts and the nature of F that $|A| < |B|$. □

By Lemma 1, every positive constant length function satisfies the ρ-set of a finite semiorder. Hence, by Theorem 2, every finite semiorder has a single-length representation.

7.3 PROBABILITY INTERVALS

Although subjective probability has historical roots in the works of Bayes (1763) and Laplace (1812), its development under the modern axiomatic idiom is due largely to Ramsey (1931), de Finetti (1937), and Savage (1954). The purpose of the present section is to show how Theorem 1 can be applied to a conjunction of the notions of comparative probability and interval orders to yield a representation for subjective probability based on intervals rather than point probabilities. The novel aspect of this application is the additivity property of probability. Other applications of Theorem 1 that involve finite additive measurement are discussed in Kraft, Pratt, and Seidenberg (1959), Adams (1965), Fishburn (1970b), and Krantz et al. (1971).

Let S be a nonempty finite set of states, and let $\mathcal{S}$ be the family of subsets of S, called events. Intuitively, we shall think of $A \prec B$ for events A and B as "A is judged to be less probable than B." It will be assumed that $(\mathcal{S}, \prec)$ is a monotonic interval order, where monotonicity means that, for all $A, B, C \in \mathcal{S}$,

$$(A \subseteq B, B \prec C) \Rightarrow A \prec C; \qquad (A \prec B, B \subseteq C) \Rightarrow A \prec C.$$

In other words, $(\subseteq)(\prec) \subseteq (\prec)$ and $(\prec)(\subseteq) \subseteq (\prec)$. This is the only property of the theorem given later that is not necessary for the representation, but it is clearly a property that reasonable judgments ought to satisfy.

Our theorem will also use the left endpoint ordering $\prec^-$, with $A \prec^- B$ if $A \sim C \prec B$ for some C. We shall say that $(\mathcal{S}, \prec)$ is *left-justified* if, for all $A_1, \ldots, A_m, B_1, \ldots, B_m$ in $\mathcal{S}$ for $m \geq 1$ such that

$$|\{i : s \in A_i\}| = |\{i : s \in B_i\}| \quad \text{for all } s \in \mathcal{S}$$

it is not true that $A_i \prec^- B_i$ for $i = 1, \ldots, m$. In other words, if each state appears equally often in the A_i and the B_i, then left-justification requires $\text{not}(A_i \prec^- B_i)$ for at least one i.

Probability Intervals

As usual, a function p on $\mathscr{S}$ is a *probability measure* if $p(\mathscr{S}) \subseteq [0,1]$, $p(S) = 1$ and, for all disjoint $A, B \in \mathscr{S}$, $p(A \cup B) = p(A) + p(B)$. We write $p(\{s\})$ as $p(s)$ for $s \in S$.

Theorem 3. *Suppose $(\mathscr{S}, \prec)$ is a monotonic interval order. Then there is a probability measure p on $\mathscr{S}$ and a real function $\delta \geq 0$ on $\mathscr{S}$ such that, for all $A, B \in \mathscr{S}$,*

$$A \prec B \Leftrightarrow p(A) + \delta(A) < p(B),$$

if and only if $(\mathscr{S}, \prec)$ is left-justified.

Proof. The proof that left-justification is necessary for the representation is straightforward and will be left to the reader. To facilitate the use of Theorem 1 in the converse proof, we remark that if the desired representation holds, then slight perturbations show that it holds with $p(s) > 0$ for all $s \in S$, $\delta > 0$, and $p(A) + \delta(A) > p(B)$ whenever $A \sim B$. Thus, given that $(\mathscr{S}, \prec)$ is a left-justified interval order, we are to show that there are real functions p on S and δ on $\mathscr{S}$ such that, for all $s \in S$ and all $A, B \in \mathscr{S}$,

$$p(s) > 0,$$

$$\Sigma_B p(s) - \Sigma_A p(s) - \delta(A) > 0 \quad \text{whenever } A \prec B,$$

$$\Sigma_B p(s) - \Sigma_A p(s) + \delta(B) > 0 \quad \text{whenever } A \sim B.$$

Note that $\delta(A) > 0$ is assured by the final inequalities when $A = B$. If $\Sigma p(s) \neq 1$, simply divide everything by this sum to get a representation for the original problem.

Let $S = \{s_1, s_2, \ldots, s_n\}$ and let w be the $(n + 2^n)$-dimensional vector for p and δ defined by

$$w = (p(s_1), \ldots, p(s_n), \delta(\varnothing), \delta(s_1), \ldots, \delta(s_n), \delta(\{s_1, s_2\}), \ldots, \delta(S)).$$

The preceding system of inequalities can then be written as

$$w \cdot a^k > 0 \quad \text{for } k = 1, \ldots, K,$$

where each a^k is an $(n + 2^n)$-dimensional vector with components in $\{-1, 0, 1\}$ that corresponds to one of the foregoing inequalities, and K is the number of inequalities.

Suppose there is no w solution to this system. Then, by Theorem 1, there are nonnegative integers $r_1, \ldots, r_K$, at least one of which is positive, such that

$$\sum_{k=1}^{K} r_k a_i^k = 0 \quad \text{for } i = 1, \ldots, n, n+1, \ldots, n + 2^n.$$

These equations correspond to $r = \Sigma r_k$ inequalities in the original (p, δ)

system, with replications for the $r_k \geq 2$. It follows from the equations in r_k for the $i > n$ that, for every $A \in \mathscr{S}$, the number of inequalities out of the total of r that have $-\delta(A)$ in an original correspondent [$A \prec B$ for some B] equals the number that have $+\delta(A)$ [$B \sim A$ for some B]. Along with the equations in r_k for $i \leq n$, this means that there is an $m \geq 1$ such that there are $A_1, \ldots, A_m, B_1, \ldots, B_m, A_{m+1}, \ldots, A_{2m}, B_{m+1}, \ldots, B_{2m}$ in $\mathscr{S}$ and $s^1, s^2, \ldots, s^t$ in S [for $r_k > 0$, if any, that correspond to the $p(s_i) > 0$] that satisfy the following:

$$A_j \prec B_j \quad \text{for } j \leq m,$$
$$A_j \sim B_j \quad \text{for } m < j \leq 2m,$$
$$A_j = B_{j+m} \quad \text{for } j \leq m,$$
$$\varnothing \sim \{s^k\} \quad \text{for } k = 1, \ldots, t,$$

and

$$\sum_{j=1}^{2m} A_j = \sum_{j=1}^{2m} B_j + \sum_{k=1}^{t} \{s^j\},$$

where equality means that each state appears the same number of times on each side. By the definition of $\prec^-$, the first three lines in the preceding display collapse to

$$A_{j+m} \prec^- B_j \quad \text{for } j \leq m,$$

and the preceding equation reduces to

$$\sum_{j=1}^{m} A_{j+m} = \sum_{j=1}^{m} B_j + \sum_{k=1}^{t} \{s^j\}.$$

It follows from monotonicity that $(s \in A, A \prec^- B) \Rightarrow A \setminus \{s\} \prec^- B$. Hence we can remove the s^j from the A_{j+m}, denote the A_{j+m} thus reduced by A'_{j+m}, and arrive at

$$\sum_{j=1}^{m} A'_{j+m} = \sum_{j=1}^{m} B_j \quad \text{and} \quad A'_{j+m} \prec^- B_j \quad \text{for } j = 1, \ldots, m.$$

But this contradicts the hypothesis that $(\mathscr{S}, \prec)$ is left-justified. Hence we conclude that there is a w solution to $\{w \cdot a^k > 0 \colon k = 1, \ldots, K\}$, and the proof is complete. □

7.4 CANTOR'S THEOREM

We now turn to the matter of representing arbitrary interval orders by families of real intervals. A fundamental element in the study of real representations is the following theorem, due essentially to Cantor (1895). We say that a subset Y of X is *order dense* in a linearly ordered set $(X, \prec)$ if, for all $x, y \in X \setminus Y$, $x \prec y$ implies $x \prec z \prec y$ for some $z \in Y$. Thus, Y is order dense in $(X, \prec)$ if something in Y is between every two distinct points not in Y.

Theorem 4 (Cantor). *Suppose $(X, \prec)$ is a linearly ordered set. Then $(X, \prec)$ is isomorphic to a restriction of $(\mathbb{R}, <)$ if and only if X includes a countable subset that is order dense in $(X, \prec)$.*

Recent proofs appear in Fishburn (1970b) and Krantz et al. (1971). Following the proof, we shall note obvious ways that Theorem 4 applies to weak orders and interval orders. Less obvious applications for interval orders and semiorders are discussed in the next two sections.

Proof. Suppose first that Y is a countable subset of X that is order dense in $(X, \prec)$. Let $A = \{(x, y): x \prec y$ and there is no $z \in Y$ such that $x \prec z \prec y\}$. If $(x, y) \in A$ then either $x \in Y$ or $y \in Y$, so A is countable. Hence Y along with all elements that appear in A, plus the first and last elements (if present) in $(X, \prec)$, forms a countable set Y^*. By Theorem 1.4 (or Remark 2 thereafter), there exists $g: Y^* \to \mathbb{R}$ such that, for all $x, y \in Y^*$, $x \prec y \Leftrightarrow g(x) < g(y)$. Define $g(z)$ for $z \in X \setminus Y^*$ by

$$g(z) = \sup\{g(y): y \in Y^* \text{ and } y \prec z\}.$$

It is not hard to see that, for all $x, y \in X$, $x \prec y \Leftrightarrow g(x) < g(y)$. Hence $(X, \prec)$ is isomorphic to some restriction of $(\mathbb{R}, <)$.

Conversely, suppose $g: X \to \mathbb{R}$ is such that, for all $x, y \in X$, $x \prec y \Leftrightarrow g(x) < g(y)$. Let $\mathscr{R}$ denote the countable set of closed intervals in $\mathbb{R}$ with distinct, rational endpoints. For each interval in $\mathscr{R}$ that contains a $g(y)$, select one such y and let B be the countable set of points in X thus selected. Also let

$$C = \{(x, y): x \prec y \text{ and } x \prec b \prec y \text{ for no } b \in B\}.$$

If $(x, y) \in C$, then it follows from the definition of B that $x \prec z \prec y$ for no $z \in X$. Hence no two open intervals $(g(x), g(y))$ for $(x, y) \in C$ intersect, and C is countable. Finally, let Y equal B in union with all points that appear in C. Then Y is countable, and it is order dense in $(X, \prec)$ since, if $x \prec y$ and $x, y \in X \setminus Y$, then $x \prec z \prec y$ for some $z \in Y$. □

The following corollary is an easy consequence of Theorem 4 in view of Theorem 1.2.

Corollary 1. *Suppose $(X, \prec)$ is a related set. Then there exists $g: X \to \mathbb{R}$ such that, for all $x, y \in X$,*

$$x \prec y \Leftrightarrow g(x) < g(y),$$

if and only if $(X, \prec)$ is a weakly ordered set and $X/\sim$ includes a countable subset that is order dense in $(X/\sim, \prec)$.

In the next corollary, $\prec_0$ with symmetric complement $\sim_0$ is the conjoint weak order on $X^- \cup X^+$ of interval order $(X, \prec)$ as defined prior to Theorem 2.3.

Corollary 2. *Suppose $(X, \prec)$ is an interval order. Then $(X, \prec)$ has a real representation F if $(X^- \cup X^+)/\sim_0$ includes a countable subset that is order dense in $((X^- \cup X^+)/\sim_0, \prec_0)$.*

By the definitions at the end of Section 2.4, F in Corollary 2 is presumed to be a mapping from X into *closed* real intervals. The possibility of real representations that use open or half-open intervals as well as closed intervals is discussed in Section 7.6.

The proof of Corollary 2 is very similar to the proof of Theorem 2.6. Under the conditions of the corollary, assume that $(X, \prec)$ is an interval order such that $(X^- \cup X^+)/\sim_0$ includes a countable subset that is order dense in the linearly ordered (Theorems 1.2 and 2.3) set $((X^- \cup X^+)/\sim_0, \prec_0)$. Let g be a $\prec_0$-preserving real function on $(X^- \cup X^+)/\sim_0$ guaranteed by Theorem 4. Then let

$$F(x) = [g([x^-]), g([x^+])],$$

recalling that $(X^- \cup X^+)/\sim_0 = (X^-/\sim^-) \cup (X^+/\sim^+)$ by Theorem 2.3 and that $[x^-]$ is the class in $X^-/\sim^-$ that contains x, and similarly for $[x^+]$. We then get $x \prec y \Leftrightarrow x^+ \prec_0 y^- \Leftrightarrow g([x^+]) < g([y^-]) \Leftrightarrow F(x) < F(y)$.

7.5 CLOSED-INTERVAL REPRESENTATIONS

This section identifies conditions for interval orders and semiorders that are necessary and sufficient for representations by closed real intervals in accord with our usual practice of viewing representations in terms of closed intervals in linearly ordered sets. However, because other types of real intervals will be considered in the next section, we shall be explicit about the types of real intervals that are used. For the present section we shall say that $(X, \prec)$ has a *closed real representation* if and only if there are functions $f, g: X \to \mathbb{R}$ with $f \leq g$ such that, for all $x, y \in X$,

$$x \prec y \Leftrightarrow g(x) < f(y).$$

This is tantamount to taking $F(x) = [f(x), g(x)]$, with $x \prec y \Leftrightarrow F(x) < F(y)$, and we shall use either F or $[f, g]$ to denote such a representation. The possibility of degenerate intervals will be allowed, with $f(x) = g(x)$, since, for example, $(\mathbb{R}, <)$ has a closed real representation only if we admit zero-length intervals.

Closed-Interval Representations

The notation of Section 2.2 ($\prec^-$, $\prec^+$, X^-, X^+, $\prec_0$, $\sim_0$, and so forth) will be used, as in the last part of the preceding section, and it may be helpful to review Sections 2.2 and 2.4 before proceeding. In addition, as in Section 3.4, we shall refer to $x \in X$ as a *simplicial point* if it is a simplicial point in the graph $(X, \sim)$. The set of all simplicial points for interval order $(X, \prec)$ is denoted by S:

$$x \in S \Leftrightarrow [\text{for all } y, z \in X, (x \sim y, x \sim z) \Rightarrow y \sim z].$$

It is evident from Theorem 4 that an interval order $(X, \prec)$ has a closed real representation only if each of $(X/\sim^-, \prec^-)$ and $(X/\sim^+, \prec^+)$ includes a countable order dense subset. However, as shown by the following example, more is needed.

Example 1. Let $X = \mathbb{R}$ with $x \prec y$ if and only if $x + 1 \leq y$. Then $(X, \prec)$ is a semiorder with $\prec^- = \prec^+ = \prec$, so the rational numbers are order dense in $(X, \prec^-)$ and in $(X, \prec^+)$. Moreover, $(X, \prec)$ can be represented by half-open intervals $[x, x + 1)$, since $x \prec y \Leftrightarrow [x, x + 1) < [y, y + 1)$. However, $(X, \prec)$ does not have a closed real representation, for suppose to the contrary that $[f, g]$ is such a representation. Then, for all real x and all $0 < \delta < 1$, we require

$$g(x) < f(x + 1) \leq g(x + \delta),$$

so that g is everywhere strictly increasing and discontinuous with an uncountable number of jumps, which is impossible.

To account for such jumps, or gaps in $\prec_0$ between an $[x^+]$ and a succeeding $[y^-]$, let

$G = \{(A, B): A \in X^+/\sim^+, B \in X^-/\sim^-, A \prec_0 B$, and there is no $C \in (X^+/\sim^+) \cup (X^-/\sim^-)$ with $A \prec_0 C \prec_0 B\}$.

If G is uncountable, then $(X, \prec)$ has no closed real representation. However, because of the countable order denseness conditions for $\prec^-$ and $\prec^+$, we need only account for the residue of pairs in G that involve no simplicial points. Hence, let

$G_0 = \{(A, B) \in G:$ there is no $x \in S$ such that either $x^+ \in A$ or $x^- \in B\}$.

The role of simplicial points is further clarified by the second of our two theorems for interval orders.

Theorem 5. *Suppose $(X, \prec)$ is an interval order. Then $(X, \prec)$ has a closed real representation if and only if each of $(X, \sim^-, \prec^-)$ and $(X/\sim^+, \prec^+)$ has a countable order dense subset and G_0 is countable.*

Theorem 6. *Suppose* $(X, \prec)$ *is an interval order that has a closed real representation. Then it has a representation* $[f, g]$ *such that, for all* $x, y \in X$,

$$f(x) = g(x) \Leftrightarrow x \in S,$$
$$f(x) < f(y) \Leftrightarrow x \prec^- y,$$
$$g(x) < g(y) \Leftrightarrow x \prec^+ y.$$

Hence simplicial points, but no others, can always be mapped into one-point intervals in a closed real representation. This applies also to interval graphs. In particular, an interval graph has a closed real representation if and only if some agreeing interval order has a closed real representation, and when such an interval order exists, all simplicial points of the graph can be mapped into degenerate intervals.

The proofs of Theorems 5 and 6 will be given after we present our results for semiorders.

When $(X, \prec)$ is a semiorder, we know by Theorem 2.2 that $\prec^- \cup \prec^+$ is a weak order on X, with $(\prec^- \cup \prec^+) = \prec^- = \prec^+$ by the definitions in Section 5.4. Moreover, since $\approx = (\sim^- \cap \sim^+)$, the symmetric complement of $\prec^- \cup \prec^+$ is the equivalence relation $\approx$. The first of our two semiorder theorems compares to Theorem 5. We do not state a separate version of Theorem 6 for semiorders since it would add nothing to Theorem 6. Instead, we note that the uniformity feature in the semiorder representation of Theorem 2.7 applies also to closed real representations.

Theorem 7. *Suppose* $(X, \prec)$ *is a semiorder. Then* $(X, \prec)$ *has a closed real representation if and only if* $(X/\approx, \prec^- \cup \prec^+)$ *has a countable order dense subset and* G_0 *is countable.*

Theorem 8. *Suppose* $(X, \prec)$ *is a semiorder that has a closed real representation. Then it has a representation* $[f, g]$ *such that, for all* $x, y \in X$,

$$f(x) < f(y) \Leftrightarrow g(x) < g(y).$$

The uniform representation of Theorem 8 is not necessarily compatible with the representation of Theorem 6 where simplicial points are mapped into one-point intervals.

One other observation is pertinent before we begin our proofs. By Theorem 3.10, every connected indifference graph is uniquely orderable. Since different components of an indifference graph will occupy different parts of $(\mathbb{R}, <)$ in a closed real representation, it follows that an indifference graph has a closed real representation (that can map simplicial points into degenerate intervals, or that can satisfy the uniformity property of Theorem 8) if and only if every semiorder that agrees with the indifference graph has a closed real representa-

Closed-Interval Representations

tion. A similar remark applies of course to the more general real representations discussed in the next section.

For convenience in the proofs that follow, let

$$\prec^* = \prec^- \cup \prec^+$$

and define I on X by

xIy if there are $z, w \in X$ with $z \prec w$ and $\{z, w\} \sim \{x, y\}$.

The union of $\prec^*$ and I will be used to denote that the left endpoint of one interval must precede the right endpoint of another (possibly the same under I) interval. The possibilities for $x(\prec^* \cup I)y$ are illustrated in Fig. 7.2. In each of the cases shown, a representation $[f, g]$ requires $f(x) < g(y)$. If xIx, then x cannot be a simplicial point. In fact, $xIx \Leftrightarrow x \notin S$.

Proofs of Theorems 5 and 6. Assume that $(X, \prec)$ is an interval order. The necessity of the countability conditions in Theorem 5 was noted above: rigorous proofs are left to the reader. To prove that those conditions are sufficient for a closed real representation, we define a binary relation $\prec_a$ on $X^- \cup X^+$ by

1. $x^- \prec_a y^- \Leftrightarrow x \prec^- y$.
2. $x^+ \prec_a y^+ \Leftrightarrow x \prec^+ y$.
3. $x^+ \prec_a y^- \Leftrightarrow x \prec y$.
4. $x^- \prec_a y^+ \Leftrightarrow x \prec^* y$ or xIy.

This is the same as the definition of the conjoint weak order $\prec_0$ preceding Theorem 2.3, except for part 4, which shows that $\prec_a \subseteq \prec_0$ (hence $\prec_a$ is asymmetric), or $\sim_0 \subseteq \sim_a$. The two differ only when $x^- \sim_a y^+$ and $x^- \prec_0 y^+$, that is, $(x \sim y, y \precsim^- x, y \precsim^+ x, \text{not}(xIy))$, and it is easily seen that this occurs if and only if there is no $c \in X^- \cup X^+$ such that $x^- \prec_0 c \prec_0 y^+$.

We shall argue that $\prec_a$ is a weak order on $X^- \cup X^+$ and that the countability conditions of Theorem 5 imply that $(X^- \cup X^+)/\sim_a$ has a

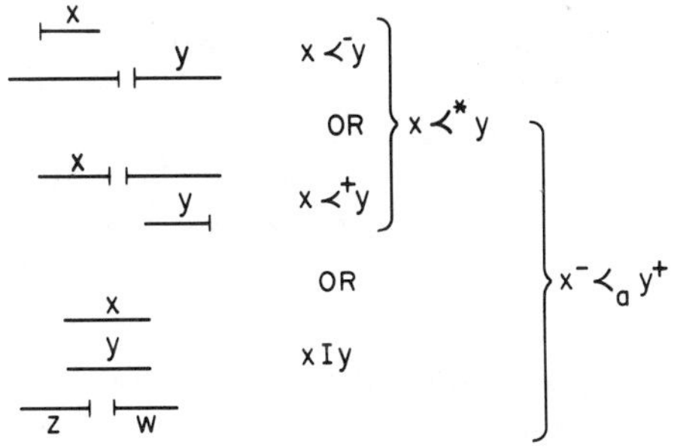

Figure 7.2 Left end of x before right end of y.

countable subset that is order dense in $((X^- \cup X^+)/\sim_a, \prec_a)$, where $\sim_a = sc(\prec_a)$. Suppose these things are true. Then, by Theorem 4, or Corollary 1, there exists $h: (X^- \cup X^+) \to \mathbb{R}$ such that, for all $c, d \in X^- \cup X^+$,

$$c \prec_a d \Leftrightarrow h(c) < h(d).$$

Let $f(x) = h(x^-)$ and $g(x) = h(x^+)$ for all $x \in X$. It is easily verified that $[f, g]$ is a closed real representation of $(X, \prec)$ which satisfies the properties in Theorem 6. Hence, to complete the proofs of Theorems 5 and 6, it remains to establish the weak order and countable order dense subset assertions.

We indicate why $(X^- \cup X^+, \prec_a)$ is a weakly ordered set without giving all the details. Recall from Theorem 2.3 that $\prec_0$ on $X^- \cup X^+$ is a weak order and that the equivalence classes in $(X^- \cup X^+)/\sim_0 = (X^-/\sim^-) \cup (X^+/\sim^+)$ "alternate" between $X^-/\sim^-$ and $X^+/\sim^+$. In addition, note that the only difference between $\prec_0$ and $\prec_a$ occurs when

$$x^- \sim_a y^+ \quad \text{and} \quad x^- \prec_0 y^+,$$

which obtains precisely when $[x^-]$ immediately precedes $[y^+]$ in the linearly ordered set $((X^- \cup X^+)/\sim_0, \prec_0)$. It then follows that $((X^- \cup X^+)/\sim_a, \prec_a)$ arises from $((X^- \cup X^+)/\sim_0, \prec_0)$ by combining classes $[x^-]$ and $[y^+]$ for $\prec_0$ into a single class $[x^-] \cup [y^+]$ for $\prec_a$ whenever $[x^-]$ immediately precedes $[y^+]$ under $\prec_0$.

Assume now that C^- is a countable order-dense subset in $(X^-/\sim^-, \prec^-)$, C^+ is a countable order-dense subset in $(X^+/\sim^+, \prec^+)$, and G_0 is countable. Let C be the set of all equivalence classes in $(X^- \cup X^+)/\sim_a$ that include something in C^- or in C^+ or in an ordered pair of G_0. Then C is countable. To prove that it is order dense in $((X^- \cup X^+)/\sim_a, \prec_a)$, suppose $A, B \in [(X^- \cup X^+)/\sim_a] \setminus C$ and $A \prec_a B$. If $[x^-] \subseteq A$ and $[y^-] \subseteq B$, then $[x^-] \prec_0 [z^-] \prec_0 [y^-]$ for some $[z^-] \in C^-$, and it follows that $A \prec_a D \prec_a B$ for some D in C. A similar conclusion holds if $[x^+] \subseteq A$ and $[y^+] \subseteq B$. If $[x^-] \subseteq A$ and $[y^+] \subseteq B$, then by Theorem 2.3 and the fact that $[x^-]$ and $[y^+]$ are not adjacent in $\prec_0$, we get $[x^-] \prec_0 [z^+] \prec_0 [w^-] \prec_0 [y^+]$ for some $z, w \in X$ and, since it is assumed that A and B are not in C, it follows that $A \prec_a D \prec_a B$ for some D in C. Finally, if $[x^+] \subseteq A$ and $[y^-] \subseteq B$, then the only way to avoid the desired conclusion is to have $[x^+]$ as the immediate predecessor of $[y^-]$ under $\prec_0$, so that $([x^+], [y^-]) \in G$. Since A and B are assumed not to be in C, this means that for some $s \in S$ either $s^+ \in [x^+]$ or $s^- \in [y^-]$. However, either possibility yields a contradiction. For example, if $s^+ \in [x^+]$ then, since $s^- \sim_a s^+$ (true for every simplicial point), $[s^-] \subseteq A$ and therefore $[s^-]$ and $[y^-]$ are adjacent in $(X^-/\sim^-, \prec^-)$ but neither is in C^-, which contradicts order denseness for C^-. $\square$

Proof of Theorem 7. Suppose $(X, \prec)$ is a semiorder, so that $\prec^*$ (the same as $<^-$ and $<^+$) is a weak order on X with symmetric complement $\approx$. Given that C is a countable order-dense subset of $(X/\approx, \prec^*)$, C is also order dense in each of $(X/\sim^-, \prec^-)$ and $(X/\sim^+, \prec^+)$, so the sufficiency proof of Theorem 5 serves also for Theorem 7. The only other thing to check is the necessity of a countable order-dense subset in $(X/\approx, \prec^*)$. Suppose $(X, \prec)$ has a closed real representation $[f, g]$. We can assume that $[f, g]$ satisfies the conclusions of Theorem 6. In particular, $f(x) < f(y) \Leftrightarrow x \prec^- y$, and $g(x) < g(y) \Leftrightarrow x \prec^+ y$, so

$$x \prec^* y \Leftrightarrow f(x) + g(x) < f(y) + g(y).$$

Then, by Corollary 1, $(X/\approx, \prec^*)$ has a countable order-dense subset. □

Proof of Theorem 8. Since this proof is similar in many ways to the sufficiency proof of Theorem 5, we omit most of the details.

Assume that $(X, \prec)$ is a semiorder that has a closed real representation. Since the type of representation in the conclusion of Theorem 8 differs from that in Theorem 6, we modify the definition of $\prec_a$ to account for this difference. Define $\prec_b$ on $X^- \cup X^+$ by $x^- \prec_b y^- \Leftrightarrow x \prec^* y$, $x^+ \prec_b y^+ \Leftrightarrow x \prec^* y$, $x^+ \prec_b y^- \Leftrightarrow x \prec y$,

$$x^- \prec_b x^+ \Leftrightarrow x \notin S \text{ or } [x \in S \text{ and, for some } y \in X, y \sim x \text{ and } y \neq x],$$

and, for $x \neq y$,

$$x^- \prec_b y^+ \Leftrightarrow x \prec^* y \text{ or } [x \sim y, y \prec^* x, \text{ and, for some } z \in X, \text{ either } x \prec^* z \sim y \text{ or } x \sim z \prec^* y].$$

It can then be shown that $\prec_b$ on $X^- \cup X^+$ is a weak order.

Since $(X, \prec)$ has a closed real representation, the countability conditions of Theorem 7 hold, and these can be shown to imply that there is a countable subset of $(X^- \cup X^+)/\sim_b$ that is order dense in $((X^- \cup X^+)/\sim_b, \prec_b)$. It follows from Corollary 1 or Theorem 4 that there is an $h: (X^- \cup X^+) \to \mathbb{R}$ such that, for all $c, d \in X^- \cup X^+$, $c \prec_b d \Leftrightarrow h(c) < h(d)$. The conclusion of Theorem 8 then follows when we let $f(x) = h(x^-)$ and $g(x) = h(x^+)$ for all $x \in X$. □

7.6 GENERAL INTERVAL REPRESENTATIONS

Example 1 shows that some interval orders that are not representable by closed real intervals can be represented by arbitrary real intervals. Because of this, we shall now consider real representability theorems for the general case. We say that $(X, \prec)$ has a *general real representation* if and only if there is a mapping F from X into nonempty intervals of $(\mathbb{R}, <)$ such that, for all $x, y \in X$,

$$x \prec y \Leftrightarrow F(x) < F(y).$$

Endpoint functions f and g with $f \le g$ will be used, but now any one of $[f(x), g(x)]$, $[f(x), g(x))$, $(f(x), g(x)]$, and $(f(x), g(x))$ could serve for $F(x)$.

Theorems 9 and 11 show that the countability condition on G_0 can be dropped from Theorems 5 and 7 when half-open and open intervals are allowed. However, the remaining conditions are not wholly necessary in the general case, as shown by Example 2. Necessary and sufficient conditions for $(X, \prec)$ to have a general real representation will not be presented. Despite this, Theorems 10 and 12 note features that any such representation can possess. A slightly different approach to real representations for semiorders is given in Świstak (1980).

Theorem 9. *Suppose $(X, \prec)$ is an interval order. Then $(X, \prec)$ has a general real representation if each of $(X/\sim^-, \prec^-)$ and $(X/\sim^+, \prec^+)$ has a countable order-dense subset.*

Theorem 10. *Suppose $(X, \prec)$ is an interval order that has a general real representation. Then it has such a representation that maps all simplicial points into one-point intervals.*

Theorem 11. *Suppose $(X, \prec)$ is a semiorder. Then $(X, \prec)$ has a general real representation if $(X/\approx, \prec^- \cup \prec^+)$ has a countable order-dense subset.*

Theorem 12. *Suppose $(X, \prec)$ is a semiorder that has a general real representation. Then it has such a representation with endpoint functions f and g such that, for all $x, y \in X$,*

$$f(x) = g(x) \quad \text{for every } x \in S,$$
$$f(x) < f(y) \Rightarrow g(x) \le g(y),$$
$$g(x) < g(y) \Rightarrow f(x) \le f(y).$$

Example 2. Let

$$X = \{(x, 0): x \in \mathbb{R}\} \cup \{(x, 1): x \in \mathbb{R}\},$$
$$F(x, 0) = [x, x+1) \quad \text{for all } x \in \mathbb{R},$$
$$F(x, 1) = (x, x+1] \quad \text{for all } x \in \mathbb{R},$$

and define $\prec$ by $a \prec b \Leftrightarrow F(a) < F(b)$. Then $(X, \prec)$ is a semiorder: in particular, $(x, 0) \prec^- (x, 1)$ and $(x, 0) \prec^+ (x, 1)$, with $\prec^- = \prec^+ = \prec^*$. More specifically,

$$(x, i) \prec^* (y, j) \Leftrightarrow [x < y \text{ or } (x = y, i = 0, j = 1)],$$

so in fact $\prec^*$ is a linear order on X. By definition, $(X, \prec)$ has a general real representation. However, it does not include a countable order-dense subset since every order-dense subset must contain at least one of the $\prec^*$-adjacent points $(x, 0)$ and $(x, 1)$ for every real x.

Proofs of Theorems 9 and 11. Because the countable order denseness condition of Theorem 11 implies those of Theorem 9 when $(X, \prec)$ is a semiorder, Theorem 11 is an immediate corollary of Theorem 9.

To prove Theorem 9, assume that $(X, \prec)$ is an interval order and each of $(X/\sim^-, \prec^-)$ and $(X/\sim^+, \prec^+)$ includes a countable order-dense subset, say C^- and C^+ respectively. We shall modify $\prec_a$ as defined in the proof of Theorem 5 to eliminate some of the gaps caused by ordered pairs in G_0. The modified $\prec_a$ will be denoted $\prec'_a$.

Recall that $\prec_a$ on $(X^- \cup X^+)/\sim_0 = (X^-/\sim^-) \cup (X^+/\sim^+)$ is the same as $\prec_0$ except that if $[x^-]$ immediately precedes $[y^+]$ in $\prec_0$, then $[x^-] \cup [y^+]$ becomes a single equivalence class in $(X^- \cup X^+)/\sim_a$. Therefore, with $A, B \in (X^- \cup X^+)/\sim_a$, we have x and y in X with

$$[x^+] \subseteq A, \quad [y^-] \subseteq B, \quad \text{and} \quad ([x^+], [y^-]) \in G_0$$

if and only if $[x^+]$ immediately precedes $[y^-]$ in $\prec_0$ and there is no simplicial point $s \in S$ such that $s^+ \in A$ or $s^- \in B$. Let G_a be the set of (A, B) that satisfy the conditions just described.

With $(A, B) \in G_a$ and $[x^+] \subseteq A$ and $[y^-] \subseteq B$, A immediately precedes B in $\prec_a$ on $(X^- \cup X^+)/\sim_a$. To obtain $\prec'_a$ from $\prec_a$, merge A and B into a single equivalence class for $\prec'_a$ if and only if there is *no* $t \in X$ such that either

$$([t^-] \sim_a [x^+], [y^-] \lesssim_a [t^-]) \quad \text{or} \quad ([t^-] \lesssim_a [x^+], [t^+] \sim_a [y^-]).$$

Otherwise, $\prec'_a$ is the same as $\prec_a$. By avoiding these cases for t in the mergers, we are assured that more than two successive classes in $(X^- \cup X^+)/\sim_a$ are not combined into one equivalence class in $(X^- \cup X^+)/\sim'_a$.

The number of $(A, B) \in G_a$ excluded from mergers is countable, for if the first alternative for t holds then $[t^-]$ and $[y^-]$ are adjacent in $X^-/\sim^-$ and therefore one of $[t^-]$ and $[y^-]$ must be in C^-, and if the second alternative for t holds, then $[x^+]$ and $[t^+]$ are adjacent in $X^+/\sim^+$ and hence one of them must be in C^+. Then, as in the last paragraph of the proof of Theorem 5, it follows that there is a countable subset of $(X^- \cup X^+)/\sim'_a$ that is order dense in $((X^- \cup X^+)/\sim'_a, \prec'_a)$.

Therefore, by Theorem 4, we get $h: (X^- \cup X^+) \to \mathbb{R}$ such that, for all $c, d \in X^- \cup X^+$,

$$c \prec'_a d \Leftrightarrow h(c) < h(d).$$

Given h, define $F(x)$ as the interval from $h(x^-)$ to $h(x^+)$ that is closed at its left end if and only if x^- is not involved in a merger of $\sim_a$ classes to obtain the $\sim'_a$ classes, and is closed at its right end if and only if x^+ is not involved in a merger. It then follows from the aspects of $\prec_a$ retained by $\prec_a'$ and from the exclusion of certain mergers for $(A, B) \in G_a$ by the cases for t that, for all $x, y \in X$, $x \prec y \Leftrightarrow F(x) < F(y)$. □

Proof of Theorem 10. Assume that $(X, \prec)$ is an interval order that has a general real representation F. We can assume that F is bounded since this can always be obtained by a monotonic transformation. For each point $x \in X$ let $S(x) = \{y \in X: x \sim y \text{ and } y \in S\}$ and define a new interval $G(x)$ for x by

$$G(x) = F(x) \cup [\cup \{F(y): y \in S(x)\}].$$

It is not difficult to show that G is a general real representation for $(X, \prec)$: we omit the details. Now use G to define H by

$$H(x) = G(x) \quad \text{if } x \notin S,$$

$$H(x) = \left\{ \frac{\inf G(x) + \sup G(x)}{2} \right\} \quad \text{if } x \in S.$$

Again, H can be shown to be a general real representation, and the proof is complete since H maps each simplicial point into a one-point interval. □

Proof of Theorem 12. Assume that $(X, \prec)$ is a semiorder that, in view of the preceding proof, has a general real representation that is bounded with endpoint functions f and g, and that maps simplicial points into one-point intervals. Assume also that $F(x) = F(y)$ if $x \approx y$. Note that $x \prec^- y$ requires $f(x) \leq f(y)$, and $x \prec^+ y$ requires $g(x) \leq g(y)$. We shall modify F as needed to satisfy *monotonicity*, that is,

$$f(x) < f(y) \Rightarrow g(x) \leq g(y),$$

$$g(x) < g(y) \Rightarrow f(x) \leq f(y),$$

without changing the $F(x)$ for $x \in S$. If monotonicity fails [$f(x) < f(y)$, $g(y) < g(x)$], then at least one of x and y cannot be in S.

Suppose first that monotonicity fails with $x \in S$, so that $x \sim y$ and

$$f(y) < f(x) = g(x) < g(y).$$

Then, since $x \neq y$, either $(y \prec^- x, x \sim^+ y)$ or $(x \sim^- y, x \prec^+ y)$. If $(y \prec^- x, x \sim^+ y)$, then we move the right end of $F(y)$ back to $g(x)$ and close it; if $(x \sim^- y, x \prec^+ y)$, then the left end of $F(y)$ is increased to $f(x)$ and closed. Neither change affects the validity of the representation. For example, the change in the first case would be inadmissible only if there were a $z \in X$ with $x \prec z \sim y$, which implies $x \prec^+ y$ in contradiction to $x \sim^+ y$. Note also that if $y \notin S$ then, by (2.2) of Section 2.1, there can be at most two $x \in S$ that are not equivalent and give $x \sim y$. If monotonicity is violated by two such $x \in S$ for a given $y \notin S$, then both ends of $F(y)$ will be moved.

General Interval Representations

Assume henceforth that F incorporates the modifications of the preceding paragraph. Then monotonicity fails only if there are $x, y \notin S$ such that either

$$\alpha: \quad (x \prec^- y, x \sim^+ y) \quad \text{and} \quad [f(x) < f(y), g(y) < g(x)],$$

or

$$\beta: \quad (x \sim^- y, x \prec^+ y) \quad \text{and} \quad [f(y) < f(x), g(x) < g(y)].$$

We now expand intervals as follows. For every instance of α, replace $g(y)$ by $\sup\{g(x): x \prec^- y, x \sim^+ y\}$ and open the upper end of the interval for y. Now consider β for the F just modified for every instance of α. If β holds after the changes for α, replace $f(x)$ by $\inf\{f(y): x \sim^- y, x \prec^+ y\}$ and open the lower end of the new interval for x. This is done for every instance of β.

We claim that these expansions do not violate the validity of the representation and that they do not change the previously established monotonicity for comparisons that involve simplicial points. The increase in $g(y)$ for α violates the representation if and only if there are x and t in X for which $y \prec t$, $x \prec^- y$, $x \sim^+ y$, and $f(t) < g(x)$. But $y \prec t$ and $x \prec^* y$ imply $x \prec t$, which requires $g(x) \leq f(t)$. Moreover, a change in $g(y)$ for α will affect monotonicity between y and $s \in S$ if and only if $g(y) = g(s) = f(s)$. But then, by α, $f(x) < f(s) = g(s) < g(x)$, which contradicts monotonicity between x and s as previously established. Similar calculations apply for β.

The only thing left to check is that the changes for β do not create a new instance of α. Since the g values are not changed for β, a new α will arise if and only if α holds after all changes have been made and, prior to the final changes for β, we have $f(x) = f(y)$ and $(x \sim^- z, x \prec^+ z)$ for some $z \in X$ for the first part of β so that $f(x)$ will be decreased in the β changes. But then $(x \prec^- y, x \sim^+ y)$ and $(x \sim^- z, x \prec^+ z)$, which imply $(z \prec^- y, y \prec^+ z)$, which is impossible when $(X, \prec)$ is a semiorder. □

8

Bounded Interval Orders

Throughout the rest of the book we shall concentrate on aspects of interval lengths in real representations of finite interval orders and interval graphs. The final two chapters focus on numbers of different lengths used in representations. The present chapter investigates classes of interval orders and graphs that have representations all of whose interval lengths are contained in a fixed closed real interval $[p, q]$ with $0 < p \leq q$. For example, $p = q$ identifies the class of finite semiorders, or the class of finite indifference graphs. Our main result, Theorem 4, implies that if p and q are integers, or if q/p is rational, then the class of interval orders for $[p, q]$ is characterized by a finite list of forbidden interval orders. A similar result holds for interval graphs.

8.1 BOUNDED INTERVAL LENGTHS

It is assumed throughout the rest of the book that X is a nonempty finite set. Because of this, we shall not make special reference to finiteness for X in the ensuing theorems. Moreover, all representations of interval orders $(X, \prec)$ and interval graphs $(X, \sim)$ will be of the form (f, ρ) with *strictly positive* length function ρ. For interval orders, $x \prec y \Leftrightarrow f(x) + \rho(x) < f(y)$; for interval graphs, $x \sim y \Leftrightarrow [f(x) \leq f(y) + \rho(y)$ and $f(y) \leq f(x) + \rho(x)]$. When $(X, \prec)$ is the main object of discussion, $\sim$ is to be understood as its symmetric complement. In other cases, $(X, \sim)$ will stand on its own as an interval graph.

For every pair $p \leq q$ of positive real numbers, we define classes of interval orders and interval graphs with length function images in $[p, q]$ as follows:

$$\mathscr{P}[p, q] = \{(X, \prec) : (X, \prec) \text{ is a finite interval order some}$$
$$\text{representation of which has } \rho(X) \subseteq [p, q]\},$$

$$\mathscr{I}[p, q] = \{(X, \sim) : (X, \sim) \text{ is a finite interval graph some}$$
$$\text{representation of which has } \rho(X) \subseteq [p, q]\}.$$

Since $\mathscr{P}[p, q] = \mathscr{P}[1, q/p]$ and $\mathscr{I}[p, q] = \mathscr{I}[1, q/p]$, it is only necessary to

Bounded Interval Lengths

consider the ratio q/p for these classes. However, it is sometimes convenient to be explicit about both p and q, so we shall use the more explicit form.

By Theorem 2.9, $\mathcal{P}[1,1]$ is the class of finite semiorders, and, by Theorem 3.2, $\mathcal{I}[1,1]$ is the class of finite indifference graphs. The reasons for our present focus on finite X stem from two earlier observations. Namely, from the latter part of Chapter 7, some infinite semiorders or interval orders have no real representation; and, by the remarks following Theorem 2.9, some infinite semiorders that have an (f,ρ) representation do not have a representation in which ρ is constant.

Our first result for the $\mathcal{P}$ and $\mathcal{I}$ classes identifies an invariant feature of the set of all interval orders that agree with a given interval graph. We noted previously (Corollaries 2.1 and 3.1) that magnitude is another invariant: all interval orders that agree with a finite interval graph have the same magnitude. The following theorem says that all agreeing interval orders lie in precisely the same collection of $\mathcal{P}[p,q]$ classes.

Theorem 1. *An interval graph $(X, \sim)$ is in $\mathcal{I}[p,q]$ if and only if every interval order that agrees with $(X, \sim)$ is in $\mathcal{P}[p,q]$.*

Proof. Let $(X, \sim)$ be a finite interval graph. According to Hanlon's (1982) analysis and Theorem 3.11, we can construct a rooted tree for $(X, \sim)$ as follows. The root is $(X, \sim)$. The "points" adjacent to the root are the components of $(X \setminus U, \sim)$, where U is the set of universal points in $(X, \sim)$. If $U = \emptyset$ and $(X, \sim)$ is connected, let $(X, \sim)$ be the only "point" adjacent to the root in this first level. The second-level points in the tree, each of which is adjacent to exactly one first-level point, are the maximal buried subgraphs (Section 3.6) of the components in the first level. The third-level points adjacent to a maximal buried subgraph in the second level are the components of that subgraph. The construction of the tree continues in the obvious way. Even-level points (beyond the root) are maximal buried subgraphs within components at the preceding level, and odd-level points are components in the root (minus U, level 1) or in a maximal buried subgraph at the preceding level. Each terminal point of the tree is a connected subgraph of $(X, \sim)$ that includes no buried subgraph. Figure 8.1 illustrates the construction.

Hanlon's analysis reveals that all interval orders that agree with $(X, \sim)$ can be obtained from each other by the following two operations:

1. Permutations of the order of main components (level 1), and permutations of the orders of components within buried subgraphs at odd levels ≥ 3.
2. Taking duals of orders within components or within buried subgraphs at any level.

Suppose $(X, \prec)$ is an interval order that agrees with $(X, \sim)$, that is, $\sim = sc(\prec)$, and is in $\mathcal{P}[p,q]$. Also let (f, ρ) be a representation of $(X, \prec)$ for which $\rho(X) \subseteq [p,q]$. If $U = \emptyset$ but $(X, \sim)$ has more than one component, each

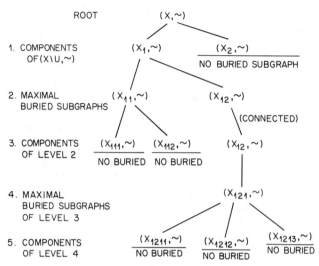

Figure 8.1 Rooted tree for an interval graph.

component's interval array in (f, ρ) can be uniformly shifted and/or dualized by a complete flip (180° rotation about the midpoint of the interval span for that component) to accommodate permutations of main components and duality within main components at level 1 without changing the symmetric complement of the resulting interval order. If $U \neq \emptyset$ (assume all $u \in U$ have the same long interval) and there is more than one component at level 1, the same changes can be made with all other intervals intersecting the universal—which is left intact—if, within each component, the left endpoints of leftmost intervals and the right endpoints of rightmost intervals of the component (see A_1 and B_m in Section 2.3) are moved right and left respectively as far as possible without violating the minimum length p or destroying an intersection of overlapping ($\sim$) intervals. After these adjustments, which may be needed to prevent a complete flip of an end component from placing an interval wholly outside of the universal, flips of components about the midpoints of their interval spans can be made for duality.

Similar procedures are followed for permutations of components within buried subgraphs (by definition, a buried subgraph has no universal point) and flips of inner components and buried subgraphs. To illustrate, consider a buried subgraph B, which can have several components, and points r, s, and t in its K-set $K(B)$, as shown in Fig. 8.2. If the intersection of the intervals for $K(B)$ is handled like the universal interval in the preceding paragraph, and the ends of components in B are trimmed as before to preserve p and needed intersections, then permutations and/or flips about span midpoints can be made without changing the symmetric complement of the resulting interval order. ◻

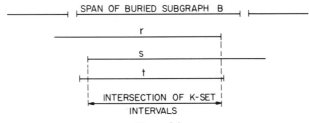

Figure 8.2

Theorem 1 will be used in Section 8.3 to prove that $\mathscr{I}[p,q]$ for rational q/p is characterized by a finite list of forbidden subgraphs. We shall note also that this is not the case when q/p is irrational.

Apart from a few remarks on the irrational ratios, ensuing sections focus on $\mathscr{P}[p,q]$ and $\mathscr{I}[p,q]$ when q/p is rational or, equivalently, when p and q are integers. The simplest integral cases are discussed in the next section. The general rational case is presented in Section 8.3. Later sections of the chapter are devoted to the proof of the main theorem, Theorem 4.

8.2 UNITARY CLASSES

We shall say that $\mathscr{P}[p,q]$ or $\mathscr{I}[p,q]$ is a *unitary class* if q/p is an integer. This section observes that unitary classes have characterizations that are nearly as elementary as the characterizations of the simplest unitary classes $\mathscr{P}[1,1]$ and $\mathscr{I}[1,1]$, for semiorders and indifference graphs respectively.

To motivate the results that follow, consider $\mathscr{P}[1,3]$ and $\mathscr{I}[1,3]$. Suppose $(X, \prec)$ is in $\mathscr{P}[1,3]$, and let (f, ρ) be a representation of $(X, \prec)$ such that $1 \leq \rho(x) \leq 3$ for every $x \in X$. Then the composition $(\sim)(\prec^4)$ is included in $\prec$, for $x \sim a \prec b \prec c \prec d \prec y$ [i.e., $x(\sim)(\prec^4)y$] implies

$$f(x) \leq f(a) + \rho(a)$$
$$f(a) + \rho(a) < f(b)$$
$$f(b) + \rho(b) < f(c)$$
$$f(c) + \rho(c) < f(d)$$
$$f(d) + \rho(d) < f(y),$$

and addition and cancellation give $f(x) + [\rho(b) + \rho(c) + \rho(d)] < f(y)$, which, in conjunction with $\rho(x) \leq \rho(b) + \rho(c) + \rho(d)$, implies $f(x) + \rho(x) < f(y)$, or $x \prec y$. Thus, $(\sim)(\prec^4) \subseteq \prec$ is necessary for membership in $\mathscr{P}[1,3]$. It also turns out to be sufficient, given that $(X, \prec)$ is a finite interval order.

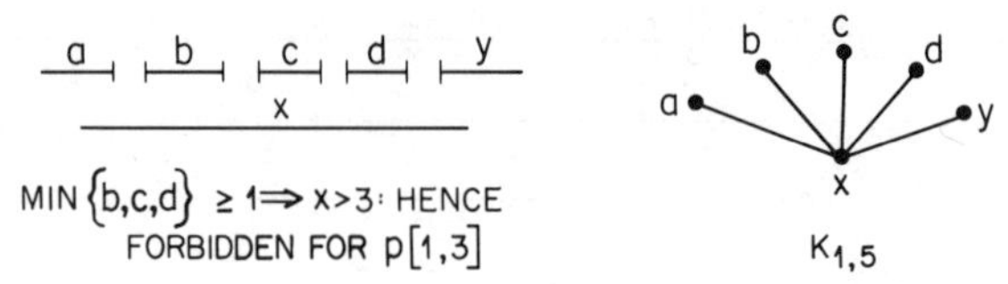

Figure 8.3

The only way that $(\sim)(\prec^4) \subseteq \prec$ can be violated is shown on the left of Fig. 8.3, with $x \sim y$ rather than $x \prec y$. The symmetric-complement graph for this violation is the bipartite graph $K_{1,5}$ shown on the right of the figure. According to Theorem 1, $K_{1,5}$ is not in $\mathcal{I}[1,3]$. In fact, an interval graph $(X, \sim)$ is in $\mathcal{I}[1,3]$ if and only if it has no subgraph isomorphic to $K_{1,5}$.

Theorem 2. *Suppose $(X, \prec)$ is an interval order and q is a positive integer. Then $(X, \prec) \in \mathcal{P}[1, q]$ if and only if $(\sim)(\prec^{q+1}) \subseteq \prec$.*

Theorem 3. *Suppose $(X, \sim)$ is an interval graph and q is a positive integer. Then $(X, \sim) \in \mathcal{I}[1, q]$ if and only if no subgraph of $(X, \sim)$ is isomorphic to $K_{1, q+2}$.*

The sufficiency proof of Theorem 2 is embedded in the sufficiency proof of Theorem 4: see Section 8.6. We note here how Theorem 3 follows from Theorems 1 and 2.

Proof of Theorem 3. Given the hypotheses of Theorem 3, suppose first that $(X, \sim)$ has no subgraph isomorphic to $K_{1, q+2}$. Then it follows easily that every interval order that agrees with $(X, \sim)$ satisfies $(\sim)(\prec^{q+1}) \subseteq \prec$ and is therefore in $\mathcal{P}[1, q]$ by Theorem 2. Hence $(X, \sim) \in \mathcal{I}[1, q]$. Conversely, if $(X, \sim)$ has a subgraph isomorphic to $K_{1, q+2}$, then some interval order that agrees with $(X, \sim)$ violates $(\sim)(\prec^{q+1}) \subseteq \prec$, so some such order is not in $\mathcal{P}[1, q]$. Then, by Theorem 1, $(X, \sim)$ is not in $\mathcal{I}[1, q]$. □

8.3 LENGTH-BOUNDED INTERVAL ORDERS

We now present conditions for $(X, \prec)$ that are necessary and sufficient for membership in $\mathcal{P}[p, q]$ whenever p and q are integers, or q/p is rational. It is then shown that these conditions, which say that certain compositions of $\sim$ and $\prec$ are included in $\prec$, imply that $\mathcal{P}[p, q]$ and $\mathcal{I}[p, q]$ for integral p and q can be characterized by finite sets of forbidden interval orders and interval graphs. We also note that this is not true when q/p is irrational.

It will be convenient in the rational case to assume that p and q are integers and either $p = 1$ (unitary) or $p > 1$ and p and q are relatively prime. The simplest example of the latter case is $\mathcal{P}[2, 3]$. Arguments similar to those in the

Length-Bounded Interval Orders

second paragraph of the preceding section show that

$$(\sim^2)(\prec^4) \subseteq \prec \quad \text{and} \quad (\prec^4)(\sim^2) \subseteq \prec$$

are necessary for membership in $\mathscr{P}[2,3]$. However, these two composition inclusions are not sufficient since $\prec^4$ is empty for the interval order pictured in Fig. 8.4, yet that order is not in $\mathscr{P}[2,3]$. In particular if x_1 through x_8 have lengths in $[2,3]$, then x_9 must be longer than 3 if it is to intersect x_1 and x_8. The order in Fig. 8.4 violates

$$(\prec^3)(\sim^2)(\prec^2)(\sim) \subseteq \prec,$$

which is necessary for membership in $\mathscr{P}[2,3]$. In the case of $[p,q] = [2,3]$, our main theorem shows that the composition inclusions that characterize $(X, \prec) \in \mathscr{P}[2,3]$ are never more complex than the one just displayed. Somewhat more involved composition inclusions are needed for other cases.

The compositions used in the inclusion conditions consist of alternating $\prec$ and $\sim$ compositions, which combine into one overall composition of the form

$$\prec^{\alpha_1} \sim^{\beta_1} \cdots \prec^{\alpha_n} \sim^{\beta_n} \quad \text{or} \quad \sim^{\beta_n} \prec^{\alpha_n} \cdots \sim^{\beta_1} \prec^{\alpha_1}.$$

For example, $x(\prec^4 \sim^2 \prec^3 \sim^5 \prec^2 \sim^1)y$ if and only if there are a, b, c, d, and e in X such that

$$x \prec^4 a, a \sim^2 b, b \prec^3 c, c \sim^5 d, d \prec^2 e, e \sim y;$$

$x \prec^4 a$ if $x \prec a_1 \prec a_2 \prec a_3 \prec a$ for some a_1, a_2, a_3; $a \sim^2 b$ if $a \sim b_1 \sim b$ for some b_1; and so forth. In other words, each overall composition alternates chains in $(X, \prec)$ with paths in its symmetric-complement graph $(X, \sim)$ and will sometimes be referred to as a chain-path composition. The superscript α_i on $\prec$ for a chain is the height of the chain (one less than the number of points in the chain), and the superscript β_i on $\sim$ for a path is the length of the path (one less than the number of terms in the path). Paths but not chains can have repeating terms, such as $a \sim b \sim b$ for $a \sim^2 b$.

The general chain-path composition inclusion conditions needed for our main theorem are specified by the following axiom scheme, which applies to integers $0 < p \leq q$ that are relatively prime.

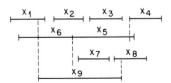

Figure 8.4 $x_1 \prec x_2 \prec x_3 \prec x_4 \sim x_5 \sim x_6 \prec x_7 \prec x_8 \sim x_9 \sim x_1$, or $x_1(\prec^3 \sim^2 \prec^2 \sim) x_9 \sim x_1$.

Axiom $A[p,q]_n$. If $(\alpha_1, \beta_1, \ldots, \alpha_n, \beta_n) \geq (2, 2, \ldots, 2, 1)$, $\sum_1^n \alpha_i = q + n$ and $\sum_1^n \beta_i = p + n - 1$, then

$$\prec^{\alpha_1} \sim^{\beta_1} \cdots \prec^{\alpha_n} \sim^{\beta_n} \subseteq \prec$$

and

$$\sim^{\beta_n} \prec^{\alpha_n} \cdots \sim^{\beta_1} \prec^{\alpha_1} \subseteq \prec.$$

Axiom $A[p,q]_1$ says that $\prec^{q+1} \sim^p \subseteq \prec$ and $\sim^p \prec^{q+1} \subseteq \prec$; $A[p,q]_2$ says that $\prec^\alpha \sim^\beta \prec^\gamma \sim^\delta$ and $\sim^\delta \prec^\gamma \sim^\beta \prec^\alpha$ are included in $\prec$ whenever $(\alpha, \beta, \gamma, \delta) \geq (2, 2, 2, 1)$, $\alpha + \gamma = q + 2$, and $\beta + \delta = p + 1$; and so forth. Although $A[p,q]_n$ places no upper limit on n relative to p and q, the following theorem shows that we do not have to consider n that exceed p.

Theorem 4. *Suppose p and q are positive integers with $p \leq q$ that are relatively prime. Suppose also that $(X, \prec)$ is an interval order. Then $(X, \prec) \in \mathcal{P}[p, q]$ if and only if $(X, \prec)$ satisfies $A[p, q]_n$ for $n = 1, \ldots, p$.*

Remarks. For unitary class $\mathcal{P}[1, q]$, the theorem says that $(X, \prec) \in \mathcal{P}[1, q]$ if and only if $[\prec^{q+1} \sim^1 \cup \sim^1 \prec^{q+1}] \subseteq \prec$. As noted earlier in Theorem 2, $\sim^1 \prec^{q+1} \subseteq \prec$ suffices for this case: see Section 8.6. The sufficiency proof of Theorem 4 is presented in the next three sections. The necessity proof is given immediately below. Corollaries of Theorem 4 and a comment on irrational q/p will then conclude the present section.

Necessity Proof of Theorem 4. Suppose p and q satisfy the hypotheses of the theorem, $(X, \prec) \in \mathcal{P}[p, q]$, (f, ρ) is a representation of $(X, \prec)$ that has $\rho(X) \subseteq [p, q]$, and

$$x_1 \prec^{\alpha_1} y_1 \sim^{\beta_1} x_2 \prec^{\alpha_2} y_2 \sim^{\beta_2} \cdots x_n \prec^{\alpha_n} y_n \sim^{\beta_n} x_{n+1}$$

is a realization of $\prec^{\alpha_1} \sim^{\beta_1} \cdots \prec^{\alpha_n} \sim^{\beta_n}$ in which the α_i and β_i satisfy the hypotheses of $A[p, q]_n$. When the (f, ρ) inequalities $[f(a) + \rho(a) < f(b)$ for $a \prec b; f(a) \leq f(b) + \rho(b)$ for $a \sim b]$ for successive terms in the composition are summed and identical factors on the two sides are cancelled, we get

$$f(x_1) + \rho(x_1) + \left[\text{sum of } \sum_{i=1}^n (\alpha_i - 1) \ \rho(\cdot)\right]$$

$$< f(x_{n+1}) + \left[\text{sum of } \sum_{i=1}^{n-1} (\beta_i - 1) + \beta_n \ \rho(\cdot)\right].$$

Since $\sum(\alpha_i - 1) = q + n - n = q$ and $p \leq \rho(\cdot)$,

$$f(x_1) + \rho(x_1) + q(p) \leq f(x_1) + \rho(x_1) + [\text{sum of } \sum(\alpha_i - 1)];$$

similarly, since $\sum_1^{n-1}(\beta_i - 1) + \beta_n = p + n - 1 - (n - 1) = p$ and $\rho(\cdot) \le q$,

$$f(x_{n+1}) + \left[\text{sum of } \sum_{i=1}^{n-1}(\beta_i - 1) + \beta_n\right] \le f(x_{n+1}) + p(q).$$

Therefore $f(x_1) + \rho(x_1) + q(p) < f(x_{n+1}) + p(q)$, so $f(x_1) + \rho(x_1) < f(x_{n+1})$ and $x_1 \prec x_{n+1}$. Hence

$$\prec^{\alpha_1} \sim^{\beta_1} \ldots \prec^{\alpha_n} \sim^{\beta_n} \subseteq \prec.$$

Alternatively, assume the initial hypotheses of the preceding paragraph along with

$$x_1 \sim^{\beta_n} y_1 \prec^{\alpha_n} x_2 \sim^{\beta_{n-1}} y_2 \prec^{\alpha_{n-1}} \ldots x_n \sim^{\beta_1} y_n \prec^{\alpha_1} x_{n+1}.$$

Then summation of the successive (f, ρ) inequalities gives

$$f(x_1) + [(\alpha_n - 1) + (\alpha_{n-1} - 1) + \cdots + (\alpha_1 - 1) \; \rho(\cdot)]$$
$$< f(x_{n+1}) + [(\beta_n - 1) + \cdots + (\beta_1 - 1) \; \rho(\cdot)],$$

and the conditions on $\Sigma\alpha_i$, $\Sigma\beta_i$, and ρ yield

$$f(x_1) + [q + n - n]p < f(x_{n+1}) + [p + n - 1 - n]q,$$

or $f(x_1) + q < f(x_{n+1})$, which implies $f(x_1) + \rho(x_1) < f(x_{n+1})$ since $\rho(x_1) \le q$. Hence

$$\sim^{\beta_n} \prec^{\alpha_n} \ldots \sim^{\beta_1} \prec^{\alpha_1} \subseteq \prec,$$

and the necessity proof is complete. □

The following corollaries of Theorems 1 and 4 imply that $\mathscr{P}[p,q]$ and $\mathscr{I}[p,q]$ are each axiomatizable by a universal sentence (no existential quantifiers in prenex normal form) of first-order logic for the $[p,q]$ used in Theorem 4. See Scott and Suppes (1958) and Titiev (1972) for elaboration on the terminology of axiomatizability. For present purposes, we state the corollaries in the finite forbidden form used previously.

Corollary 1. *Suppose p and q satisfy the hypotheses of Theorem 4. Then there is a finite list of interval orders, which depends on p and q, such that any interval order $(X, \prec)$ is in $\mathscr{P}[p,q]$ if and only if it has no restriction that is isomorphic to an interval order in the list.*

Corollary 2. *Suppose p and q satisfy the hypotheses of Theorem 4. Then there is a finite list of interval graphs, which depends on p and q, such that any interval graph $(X, \sim)$ is in $\mathscr{I}[p, q]$ if and only if it has no subgraph that is isomorphic to an interval graph in the list.*

Proofs. Given the hypotheses of Corollary 1, it follows from Theorem 4 and the statement of Axiom $A[p, q]_n$ that, up to isomorphism, there are only finitely many interval orders with $n \leq p$ that violate $A[p, q]_n$ and have no point that is not involved in a composition that yields a violation. Consequently, if interval order $(X, \prec)$ is not in $\mathscr{P}[p, q]$, it must have a restriction that is isomorphic to one of the minimal violators.

Each minimal violator of $A[p, q]_n$ for $n \leq p$ corresponds to an interval graph that has the violating interval order as one of its agreeing interval orders. According to Theorem 1, such an interval graph cannot be in $\mathscr{I}[p, q]$. Let $\mathscr{L}[p, q]$ denote the set of interval graphs thus generated. Since the minimal violators of $A[p, q]_n$ for $n \leq p$ are finite in number up to isomorphism, this is true also for $\mathscr{L}[p, q]$. As just noted, if an interval graph has a subgraph isomorphic to one in $\mathscr{L}[p, q]$, then the interval graph is not in $\mathscr{I}[p, q]$. Conversely, if $(X, \sim)$ has no subgraph isomorphic to one in $\mathscr{L}[p, q]$, then no restriction of any interval order that agrees with $(X, \sim)$ can violate $A[p, q]_n$, so every interval order that agrees with $(X, \sim)$ is in $\mathscr{P}[p, q]$, and therefore $(X, \sim)$ is in $\mathscr{I}[p, q]$. This completes the proof of Corollary 2. □

Theorem 3 noted that $\mathscr{L}[p, q]$ in the preceding proof contains only $K_{1, q+2}$ when $p = 1$. When p and q are relatively prime integers with $2 \leq p < q$, the forbidden set $\mathscr{L}[p, q]$ is considerably more complex and is not easily described apart from the correspondence between its members and the minimal interval order violators of $A[p, q]_n$ for $n \leq p$. This is true even for $\mathscr{L}[2, 3]$. For example, the minimal violators of $A[2, 3]_1$ and $A[2, 3]_2$ are interval orders with nine or fewer points, including x and y, for which

$$x\left(\sim^2 \prec^4 \cup \prec^4 \sim^2 \cup \sim^1 \prec^2 \sim^2 \prec^3 \cup \prec^3 \sim^2 \prec^2 \sim^1 \right.$$
$$\left. \cup \sim^1 \prec^3 \sim^2 \prec^2 \cup \prec^2 \sim^2 \prec^3 \sim^1 \right) y$$

and $\text{not}(x \prec y)$. The symmetric-complement graphs of these minimal violators are the graphs in $\mathscr{L}[2, 3]$.

The situation presented in Corollaries 1 and 2 is very different if q/p is irrational. Consider $\mathscr{P}[1, r]$ for irrational $r > 1$. For every rational $s > r$, say $s = a/b$ with a and b relatively prime positive integers, $(X, \prec)$ is in $\mathscr{P}[1, r]$ only if it is in $\mathscr{P}[1, s]$, so $A[a, b]_n$ must hold for all $n \leq a$ if $(X, \prec)$ is to be in $\mathscr{P}[1, r]$. Since there are rational $s = a/b > r$ that are arbitrarily near to r and have arbitrarily large a, it follows that no finite set of interval orders can characterize $\mathscr{P}[1, r]$ by forbidden isomorphic restrictions. Theorem 1 then implies that there is no finite list of interval graphs such that $(X, \sim)$ is in $\mathscr{I}[1, r]$ if and only if $(X, \sim)$ has no subgraph that is isomorphic to one in the list.

8.4 LINEAR INEQUALITIES AND FORBIDDEN PICYCLES

The rest of the chapter is devoted to the sufficiency proof of Theorem 4. In the present section we shall use the linear solution theory of Theorem 7.1 to obtain a characterization of $\mathscr{P}[p, q]$ in terms of forbidden picycles. This intermediate result is stated below as Theorem 5. The next two sections then show that if an interval order satisfies $A[p, q]_n$ for $n = 1, \ldots, p$, it has no forbidden picycle; hence, by Theorem 5, it must be in $\mathscr{P}[p, q]$.

It is assumed throughout the rest of the chapter that p and q are integers with $0 < p \leq q$, and with p and q relatively prime. In addition, $\sim$ always denotes the symmetric complement of $\prec$, and $\precsim$ is the union of $\prec$ and $\sim$.

Several definitions will be needed. Generalizing from Section 1.7, we shall refer to $(x_1, x_2, \ldots, x_K, x_1)$ as a *cycle* (of length K) for interval order $(X, \prec)$ if $x_i \precsim x_{i+1}$ for $i = 1, \ldots, K - 1$, and $x_K \precsim x_1$. The cycle is *simple* if all K x_i are different. Cycles $(x_1, x_2, \ldots, x_K, x_1), (x_2, \ldots, x_K, x_1, x_2), \ldots,$ $(x_K, x_1, \ldots, x_{K-1}, x_K)$ are viewed as equivalent.

A cycle $(x_1, \ldots, x_K, x_1)$ for $(X, \prec)$ is a *picycle* (named from the use of $\prec$ for preference and $\sim$ for indifference) if at least one $x_i \precsim x_{i+1}$ is actually $x_i \prec x_{i+1}$ ($K + 1 = 1$). A picycle must also have $x_i \sim x_{i+1}$ for some i, else $(X, \prec)$ would not be a poset.

A picycle can always be arranged so that $x_1 \prec x_2$ and $x_K \sim x_1$. In this arrangement it consists of a chain ($\prec$), followed by a path ($\sim$), followed by a chain,..., and ending in a path. A picycle $\mathscr{C}$ that is composed of n chains of lengths $\alpha_1, \alpha_2, \ldots, \alpha_n$ followed alternately with n paths of lengths $\beta_1, \beta_2, \ldots, \beta_n$, may be abbreviated as

$$\mathscr{C} = x_1 \prec^{\alpha_1} y_1 \sim^{\beta_1} \cdots x_n \prec^{\alpha_n} y_n \sim^{\beta_n} x_1$$

or as

$$\mathscr{C} = \prec^{\alpha_1} \sim^{\beta_1} \cdots \prec^{\alpha_n} \sim^{\beta_n}.$$

The *index* of $\mathscr{C}$ in this form is $(\alpha_1, \beta_1, \ldots, \alpha_n, \beta_n)$, its length is $\Sigma_1^n(\alpha_i + \beta_i)$, and its $\prec$-*excess* $e(\prec)$ and $\sim$-*excess* $e(\sim)$ are defined by

$$e(\prec) = \Sigma(\alpha_i - 1) = \Sigma\alpha_i - n,$$

$$e(\sim) = \Sigma(\beta_i - 1) = \Sigma\beta_i - n.$$

Finally, we shall say that a picycle $\mathscr{C} = \prec^{\alpha_1} \sim^{\beta_1} \cdots \prec^{\alpha_n} \sim^{\beta_n}$ for which

$$pe(\prec) \geq qe(\sim)$$

is a $[p, q]$-*forbidden* picycle.

Theorem 5. *An interval order* $(X, \prec)$ *is in* $\mathscr{P}[p, q]$ *if and only if it has no* $[p, q]$-*forbidden picycle.*

It may be noted that Theorem 5 is not generally true when "simple picycle" is substituted for "picycle." This is verified by the 19-point interval order described in Fig. 8.5. With $[p, q] = [5, 9]$, the picycle

$$\mathscr{C} = x_1 \prec x_2 \prec x_3 \prec y \sim x_4 \sim x_5 \prec x_6 \prec x_7 \prec x_8 \prec x_9 \prec x_{10} \prec x_{11}$$
$$\sim x_{12} \sim y \prec x_{13} \prec x_{14} \prec x_{15} \sim x_{16} \sim x_{17} \sim x_{18} \sim x_1$$

with index $(3, 2, 6, 2, 3, 4)$ is $[5, 9]$-forbidden since $pe(\prec) = 5(9) = 9(5) = qe(\sim)$. Hence, by Theorem 5, the interval order is not in $\mathscr{P}[5, 9]$.

However, the interval order of Fig. 8.5 has no $[5, 9]$-forbidden *simple* picycle. This can be proved by showing first that every 18-point restriction of the interval order is in $\mathscr{P}[5, 9]$. It then follows from Theorem 5 that a simple picycle can have $5e(\prec) \geq 9e(\sim)$ only if it uses all 19 points. But there is no such picycle, as can be seen by closer analysis. I leave the details to the reader.

Proof of Theorem 5. The proof that $(X, \prec)$ cannot be in $\mathscr{P}[p, q]$ when it has a picycle with $pe(\succ) \geq qe(\sim)$ is straightforward and is left to the reader. To prove the converse, suppose for the moment that $(X, \prec)$ is in $\mathscr{P}[p, q]$ with $|X| = N$ and $X = \{a_1, a_2, \ldots, a_N\}$. Then, given any $\lambda > 0$, there are real valued functions f and $\rho > 0$ on X such that

$$f(a_i) + \rho(a_i) < f(a_j) \qquad \text{whenever } a_i \prec a_j,$$
$$f(a_i) \leq f(a_j) + \rho(a_j) \qquad \text{whenever } a_i \sim a_j \text{ and } i \neq j,$$
$$p\lambda \leq \rho(a_i) \leq q\lambda \qquad \text{for } i = 1, \ldots, N.$$

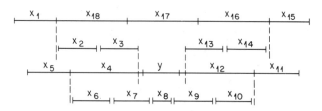

Figure 8.5 An interval order not in $\mathscr{P}[5, 9]$.

Let $w = (f(a_1), \ldots, f(a_N), \rho(a_1), \ldots, \rho(a_N), \lambda)$. Also let $a(i, j)$ for $a_i \prec a_j$ be the vector in $\mathbb{R}^{2N+1}$ that has 1 in position j, -1's in positions i and $N + i$, and 0's elsewhere, and let $a(i, j)$ for $a_i \sim a_j$ and $i \neq j$ be the vector in $\mathbb{R}^{2N+1}$ that has 1's in positions j and $N + j$, -1 in position i, and 0's elsewhere.

It follows that $(X, \prec)$ is in $\mathscr{P}[p, q]$ if and only if there is a w that satisfies

$$w \cdot (0, \ldots, 0, 1) > 0$$

$$w \cdot a(i, j) > 0 \quad \text{whenever } a_i \prec a_j$$

$$w \cdot a(i, j) \geq 0 \quad \text{whenever } a_i \sim a_j \text{ and } i \neq j,$$

$$w \cdot (0, \ldots, 0, 1_{N+i}, 0, \ldots, 0, -m) \geq 0 \quad \text{for } i = 1, \ldots, N,$$

$$w \cdot (0, \ldots, 0, -1_{N+i}, 0, \ldots, 0, n) \geq 0 \quad \text{for } i = 1, \ldots, N$$

where d_j denotes d in position j. Supposing that there are K ordered pairs in $\prec$, and a total of J inequalities in the system, we can write it as

$$w \cdot x^k > 0 \quad \text{for} \quad k = 1, \ldots, K + 1$$

$$w \cdot x^k \geq 0 \quad \text{for} \quad k = K + 2, \ldots, J,$$

where x^k is the appropriate coefficient vector from the preceding display.

Assume henceforth in this proof that $(X, \prec)$ is *not* in $\mathscr{P}[p, q]$, so there is no w solution to the system. We shall show that $(X, \prec)$ has a $[p, q]$-forbidden picycle. Given no w solution, Theorem 7.1 implies that there are nonnegative integers $r_1, \ldots, r_J$ with $r_k > 0$ for some $k \leq K + 1$ such that

$$\sum_{k=1}^{J} r_k x_i^k = 0 \quad \text{for} \quad i = 1, \ldots, 2N + 1,$$

where $x^k = (x_1^k, \ldots, x_{2N+1}^k)$. Let $r = r_1 + r_2 + \cdots + r_J$ and take r_k replicates of each basic inequality to yield the following correspondent of $\Sigma r_k x^k = \mathbf{0}$:

c_1 inequalities $\lambda > 0$,
c_2 inequalities $f(a_j) - f(a_i) - \rho(a_i) > 0 \quad (a_i \prec a_j)$,
c_3 inequalities $f(a_j) - f(a_i) + \rho(a_j) \geq 0 \quad (a_i \sim a_j)$,
c_4 inequalities $\rho(a_i) - p\lambda \geq 0$,
c_5 inequalities $-\rho(a_i) + q\lambda \geq 0$,

where $c_1 = r_1$, $c_2 = r_2 + \cdots + r_{K+1}$, $c_1 + c_2 > 0$, and $\Sigma c_i = r$. According to $\Sigma r_k x^k = \mathbf{0}$, we have the following for each $i \leq N$:

R1. The number of inequalities in the c list with $+f(a_i)$ equals the number with $-f(a_i)$.

R2. The number of inequalities with $+\rho(a_i)$ equals the number with $-\rho(a_i)$.

R3. $c_1 - pc_4 + qc_5 = 0$ [for $i = 2N + 1$].

We now work with the $\prec$ and $\sim$ pairs for c_2 and c_3, using R_1 through R_3 as needed.

Suppose first that $c_2 = 0$. Then all ρ terms for c_2 and c_3 are $+\rho(a_j)$ and, by R2, these must be balanced by $-\rho(a_j)$ terms from c_5. If $c_5 > c_3$, then $c_5 - c_3$ excess from c_5 must be balanced by terms from c_4. Hence, by R2, $c_4 = c_5 - c_3$. Then, by R3, $c_1 = pc_4 - qc_5 = p(c_5 - c_3) - qc_5 = -(q - p)c_5 - pc_3 \leq 0$, which contradicts $c_1 + c_2 > 0$. Therefore $c_2 > 0$.

Given $c_2 > 0$, arrange the $c_2 \prec$ pairs and the $c_3 \sim$ pairs in two rows as follows, where, by R1, each row is a permutation of the other:

	c_2 pairs				c_3 pairs				
row 1	x_1	x_2	$\cdots$	x_{c_2}	z_1	z_2	$\cdots$	z_{c_3}	$-f$
row 2	y_1	y_2	$\cdots$	y_{c_2}	v_1	v_2	$\cdots$	v_{c_3}	$+f$

In this array, $x_i \prec y_i$ and $z_i \sim v_i$ ($z_i \neq v_i$). We shall form picycles from the array, treating each column as distinct.

The first picycle begins with $x_1 \prec y_1$. This is followed by a $y_1 \sim v_i$ if y_1 is one of the z_i's, and by $y_1 \prec y_j$ if y_1 is not one of the z_i's. In constructing this or any other picycle, we shall always follow a $\prec$ pair by a $\sim$ pair whenever possible, and always follow a $\sim$ pair by a $\prec$ pair whenever possible, except when the current pair completes a picycle. Once a pair or column is used, it is deleted from the array. The construction of the first picycle continues until we encounter x_1 as the second member of a newly added pair *and* there are no x_1's left in the first row. As a final step, we rearrange the picycle as needed so that it begins with a $\prec$ and ends with a $\sim$.

New picycles are constructed in a similar manner until all $\prec$ pairs have been deleted. At this point, any remaining $\sim$ pairs can be formed into $\sim$ cycles since the remainder of row 2 is a permutation of the remainder of row 1.

The construction of picycles follows a $\prec$ pair by a $\sim$ pair, and a $\sim$ pair by a $\prec$ pair, whenever possible. We shall say that a *transition* occurs each time a $\sim$ pair is followed by a $\prec$ pair. Each transition has the form $a \sim b \prec c$, with $+\rho(b)$ associated with $a \sim b$ [or $f(a) \leq f(b) + \rho(b)$] and $-\rho(b)$ associated with $b \prec c$ [or $f(b) + \rho(b) < f(c)$]. Hence the ρ terms for the transitions balance out as part of R2.

We claim that the x_i in nontransitional $\prec$ columns of the array are disjoint from the v_i in nontransitional $\sim$ columns. Suppose to the contrary that

$$a \prec x_i \prec b \quad \text{is part of a picycle,}$$

$$c \sim x_i \sim d \quad \text{is part of a picycle or } \sim \text{ cycle.}$$

Reducibility Lemmas

This contradicts the method of construction. For if the picycle with $a \prec x_i \prec b$ arises in the construction first, then $a \prec x_i$ would be followed by $x_i \sim d$ (for some d); and if $c \sim x_i \sim d$ arises first, then $c \sim x_i$ would be followed by $x_i \prec b$ (for some b), and in either case we get a contradiction.

Let t be the number of transitions in the constructed picycles. By the preceding paragraph, the $(c_2 - t)$ x_i's in nontransitional $\prec$ columns have $-\rho(x_i)$ terms whose x_i are disjoint from the $(c_3 - t)$ v_j's in nontransitional $\sim$ columns, which are associated with $+\rho(v_j)$ terms. Therefore, by R2, c_4 must include $c_2 - t$ terms $(+\rho)$ to balance the $-\rho(x_i)$ ones, and c_5 must include $c_3 - t$ terms $(-\rho)$ to balance the $+\rho(v_j)$ ones. In addition, c_4 and c_5 can each have s other terms that cancel between the two: $+\rho(a)$ for c_4, $-\rho(a)$ for c_5. It follows that

$$c_4 = c_2 - t + s,$$

$$c_5 = c_3 - t + s.$$

Since R3 implies $c_1 = pc_4 - qc_5 \geq 0$, we get $p(c_2 - t + s) - q(c_3 - t + s) \geq 0$, or $p(c_2 - t) \geq q(c_3 - t) + s(q - p)$, hence

$$p(c_2 - t) \geq q(c_3 - t).$$

Finally, let $E(\prec)$ and $E(\sim)$ be the sums of the excesses $e(\prec)$ and $e(\sim)$ in the constructed picycles. Since each chain in a picycle after the first chain is associated with one transition, and each path in a picycle prior to the last path is associated with one transition, we have

$$c_2 = E(\prec) + t + u,$$

$$c_3 \geq E(\sim) + t + u,$$

where u is the number of picycles. The inequality for c_3 arises from the possibility of $\sim$ cycles left over at the end of the construction. In any event, we have $c_2 - t = E(\prec) + u$ and $c_3 - t \geq E(\sim) + u$, so $p(E(\prec) + u) \geq q(E(\sim) + u)$, and therefore

$$pE(\prec) \geq qE(\sim).$$

Hence it must be true that $pe(\succ) \geq qe(\sim)$ for some picycle which, by definition, is a $[p, q]$-forbidden picycle. □

8.5 REDUCIBILITY LEMMAS

This section establishes reducibility lemmas for picycles which will be used in the next section along with Theorem 5 to complete the sufficiency proof of

Theorem 4. For convenience, the $[p,q]$ designation will often be omitted: a forbidden picycle is a $[p,q]$-forbidden picycle, and A_n stands for $A[p,q]_n$.

A forbidden picycle $\mathscr{C} = \prec^{\alpha_1} \sim^{\beta_1} \cdots \prec^{\alpha_n} \sim^{\beta_n}$ is said to be *reducible* if some contiguous segment of $\mathscr{C}$ involving two or more $\prec$ and/or $\sim$ pairs can be replaced by one $\prec$ or $\sim$ pair that uses the first and last points in the segment so that the picycle $\mathscr{C}'$ obtained by the replacement is also forbidden. We shall let

$$h(\mathscr{C}) = \text{length of picycle } \mathscr{C}.$$

If $\mathscr{C}$ is reducible to $\mathscr{C}'$ by replacement of a segment of $\mathscr{C}$ with a single $\prec$ or $\sim$ pair, then $h(\mathscr{C}') < h(\mathscr{C})$.

Our basic reducibility lemma is

Lemma 1. *Suppose $\mathscr{C} = \prec^{\alpha_1} \sim^{\beta_1} \cdots \prec^{\alpha_N} \sim^{\beta_N}$ is a forbidden picycle. Let $\alpha_i = \alpha_{i-N}$ and $\beta_i = \beta_{i-N}$ for $N+1 \leq i \leq 2N$. Then*

(i) *$\mathscr{C}$ is reducible if $\alpha_i = 1$ or $\beta_i = 1$ for some i;*

(ii) *If $\alpha_i \geq 2$ and $\beta_i \geq 2$ for all i, and if A_n holds for a specified $n \leq N$, then $\mathscr{C}$ is reducible to $\mathscr{C}'$ with*

$$h(\mathscr{C}') = h(\mathscr{C}) - [p + q + 2(n-1)]$$

if there is an $i \in \{1, \ldots, N\}$ such that either

(a) $\alpha_i + \cdots + \alpha_{i+n-1} \geq q + n$,
$\alpha_{i+1} + \cdots + \alpha_{i+n-1} \leq q + n - 2$ if $n \geq 2$,
$\beta_i + \cdots + \beta_{i+n-1} \geq p + n$, and
$\beta_i + \cdots + \beta_{i+n-2} \leq p + n - 2$ if $n \geq 2$;

or

(b) $\beta_i + \cdots + \beta_{i+n-1} \geq p + n$,
$\beta_{i+1} + \cdots + \beta_{i+n-1} \leq p + n - 2$ if $n \geq 2$,
$\alpha_{i+1} + \cdots + \alpha_{i+n} \geq q + n$, and
$\alpha_{i+1} + \cdots + \alpha_{i+n-1} \leq q + n - 2$ if $n \geq 2$.

Proof. Given the hypotheses of the lemma, suppose first that $\beta_1 = 1$. If $N = 1$, then $\alpha_1 = 1$ is impossible (else $x \prec y \sim x$), and $\alpha_1 \geq 2$ is impossible, since $(\prec \sim \prec) \subseteq \prec$ and $\prec^2 \subseteq \prec$, which hold for all interval orders, would yield $x \prec x$. Therefore $N \geq 2$. Then, since $(\prec \sim \prec) \subseteq \prec$, $\mathscr{C}$ can be reduced to $\mathscr{C}' = \prec^{\alpha_1 + \alpha_2 - 1} \sim^{\beta_2} \ldots$, which is forbidden since it has $e'(\prec) = e(\prec)$ and $e'(\sim) = e(\sim)$, where the prime denotes $\mathscr{C}'$. It follows that $\mathscr{C}$ is reducible if $\beta_i = 1$ for any i.

Assume henceforth that $\beta_i \geq 2$ for all i. To complete the proof of conclusion (i) of the lemma, suppose some $\alpha_i = 1$. Then $pe(\prec) \geq qe(\sim)$ requires $N \geq 2$, so assume for definiteness that $\alpha_2 = 1$. Since $(\sim \prec \sim) \subseteq \precsim$ for any interval order, the $\sim \prec \sim$ part of $\sim^{\beta_1} \prec^{\alpha_2} \sim^{\beta_2}$ can be replaced by $\sim$ or $\prec$ to

Reducibility Lemmas

yield $\mathscr{C}'$. It is easily checked that either replacement gives $pe'(\prec) \geq qe'(\sim)$, so $\mathscr{C}$ is reducible.

Assume henceforth that $\alpha_i \geq 2$ for all i. Suppose for conclusion (ii) of the lemma that A_n holds for some $n \leq N$ and that the α_j and β_j satisfy the inequalities of system (a) for some $i \leq N$. For notational convenience let $i = 1$ for (a). If $n = 1$, then (a) says that $\alpha_1 \geq q + 1$ and $\beta_1 \geq p + 1$, and the $\prec^{q+1} \sim^p$ part of $\prec^{\alpha_1} \sim^{\beta_1}$ can be replaced by $\prec$ according to A_1. This changes $e(\prec)$ to $e'(\prec) = e(\prec) - q$ and $e(\sim)$ to $e'(\sim) = e(\sim) - p$, so $pe'(\prec) \geq qe'(\sim)$ and $\mathscr{C}$ is reducible with $h(\mathscr{C}') = h(\mathscr{C}) - (p + q)$. If $n \geq 2$ then (a) says that

$$\alpha_1 \geq q + n - \sum_2^n \alpha_i \geq 2 \quad \text{and} \quad \beta_n - 1 \geq p + n - 1 - \sum_1^{n-1} \beta_i \geq 1,$$

so the segment

$$\prec^{q+n-\sum_2^n \alpha_i} \sim^{\beta_1} \prec^{\alpha_2} \cdots \prec^{\alpha_n} \sim^{p+n-1-\sum_1^{n-1} \beta_i}$$

of length $p + q + 2n - 1$ can be replaced by $\prec$ according to A_n to yield picycle $\mathscr{C}'$ with

$$h(\mathscr{C}') = h(\mathscr{C}) - [p + q + 2(n - 1)].$$

Since a $\sim$ pair immediately follows the replaced segment,

$$e'(\prec) = e(\prec) - \left[\left(q + n - \sum_2^n \alpha_i - 1\right) + \sum_2^n (\alpha_i - 1)\right] = e(\prec) - q,$$

$$e'(\sim) = e(\sim) - \left[\left(p + n - 1 - \sum_1^{n-1} \beta_i\right) + \sum_1^{n-1} (\beta_i - 1)\right] = e(\sim) - p.$$

Hence $pe'(\prec) \geq qe'(\sim)$, so $\mathscr{C}$ is reducible to $\mathscr{C}'$.

The proof for system (b) is similar. □

We now use Lemma 1 to prove three subsidiary reducibility lemmas that will be needed in the next section. The proofs of these lemmas conclude this section.

Lemmas 2, 3, and 4 assume that $\mathscr{C} = \prec^{\alpha_1} \sim^{\beta_1} \cdots \prec^{\alpha_N} \sim^{\beta_N}$ is a forbidden picycle, that $\alpha_i \geq 2$ and $\beta_i \geq 2$ for all i, that $p \geq 2$ and that A_1 through A_p hold.

Lemma 2. $\mathscr{C}$ is reducible if $\alpha_i \geq q + 1$ for some i.

Lemma 3. $\mathscr{C}$ is reducible if $\beta_i \geq p + 1$ for some i.

Lemma 4. Suppose $0 \leq k \leq p - 3$, $k + 2 \leq N$, $\sum_{j=i}^{i+k}\alpha_j \leq q + k$, and $\sum_{j=i}^{i+k}\beta_j \leq p + k$ for $i = 1, \ldots, N$, where $\alpha_j = \alpha_{j-N}$ and $\beta_j = \beta_{j-N}$ for $j > N$. Then either:

(a) for each $i \leq N$,
$$\sum_{j=i}^{i+k+1} \alpha_j \leq q + k + 1 \quad \text{and} \quad \sum_{j=i}^{i+k+1} \beta_j \leq p + k + 1; \quad \text{or}$$

(b) $pe(\prec) - qe(\sim) \geq q$, in which case $\mathscr{C}$ is reducible by shortening a chain; or

(c) $\mathscr{C}$ is reducible to $\mathscr{C}'$ via Lemma 1(ii) for some $n \in \{k + 2, \ldots, \min\{N, p\}\}$, with $h(\mathscr{C}') = h(\mathscr{C}) - [p + q + 2(n - 1)]$.

Proof of Lemma 2. Assume with no loss of generality that $\alpha_1 \geq q + 1$, and suppose that $\mathscr{C}$ is not reducible. Since $\alpha_1 \geq q + 1$, Lemma 1(ii, a) requires $\beta_1 \leq p$. If $\alpha_2 \geq q + 1$, then $\beta_2 \leq p$; if $\alpha_2 \leq q$, then $\alpha_1 + \alpha_2 \geq q + 2$, and Lemma 1(ii, a) for $n = 2$ requires either $\beta_1 + \beta_2 \leq p + 1$ or $\beta_1 \geq p + 1$: since not($\beta_1 \geq p + 1$), either

$$\alpha_2 \geq q + 1 \quad \text{and} \quad \beta_2 \leq p; \quad \text{or}$$
$$\alpha_2 \leq q, \quad \alpha_1 + \alpha_2 \geq q + 2, \quad \text{and} \quad \beta_1 + \beta_2 \leq p + 1.$$

If $\alpha_3 \geq q + 1$ then $\beta_3 \leq p$; if $\alpha_3 \leq q$ and $\alpha_2 + \alpha_3 \geq q + 2$ then $\beta_2 + \beta_3 \leq p + 1$ or $\beta_2 \geq p + 1$ (which is impossible since $\beta_2 \leq p$ by the preceding sentence); if $\alpha_2 + \alpha_3 \leq q + 1$ then $\alpha_1 + \alpha_2 + \alpha_3 \geq q + 3$, and then Lemma 1(a) for $n = 3$, or A_3, requires $\beta_1 + \beta_2 + \beta_3 \leq p + 2$ or $\beta_1 + \beta_2 \geq p + 2$ (which is precluded by the fact that $\alpha_2 + \alpha_3 \leq q + 1$ implies $\alpha_2 \leq q$, hence $\alpha_1 + \alpha_2 \geq q + 2$, hence $\beta_1 + \beta_2 \leq p + 1$). Therefore either

$$\alpha_3 \geq q + 1 \quad \text{and} \quad \beta_3 \leq p; \quad \text{or}$$
$$\alpha_3 \leq q, \quad \alpha_2 + \alpha_3 \geq q + 2, \quad \text{and} \quad \beta_2 + \beta_3 \leq p + 1; \quad \text{or}$$
$$\alpha_2 + \alpha_3 \leq q + 1, \quad \alpha_1 + \alpha_2 + \alpha_3 \geq q + 3, \quad \text{and} \quad \beta_1 + \beta_2 + \beta_3 \leq p + 2.$$

The natural continuation of this procedure to any $n \leq \min\{N, p\}$ gives either

$$\alpha_n \geq q + 1 \quad \text{and} \quad \beta_n \leq p; \quad \text{or}$$
$$\alpha_n \leq q, \quad \sum_{n-1}^{n} \alpha_i \geq q + 2, \quad \text{and} \quad \sum_{n-1}^{n} \beta_i \leq p + 1; \quad \text{or}$$
$$\vdots$$
$$\sum_{2}^{n} \alpha_i \leq q + n - 2, \quad \sum_{1}^{n} \alpha_i \geq q + n, \quad \text{and} \quad \sum_{1}^{n} \beta_i \leq p + n - 1.$$

Reducibility Lemmas

Suppose $p \leq N$. Then the preceding line cannot hold at $n = p$ since it requires $\Sigma_1^p \beta_i \leq 2p - 1$, whereas $\Sigma_1^p \beta_i \geq 2p$. In addition, if $p < N$, then continuance to $n \in \{p + 1, \ldots, N\}$ gives either

$$\alpha_n \geq q + 1 \quad \text{and} \quad \beta_n \leq p; \quad \text{or}$$

$$\alpha_n \leq q, \quad \sum_{n-1}^n \alpha_i \geq q + 2, \quad \text{and} \quad \sum_{n-1}^n \beta_i \leq p + 1; \quad \text{or}$$

$$\vdots$$

$$\sum_{n-p+3}^n \alpha_i \leq p + q - 3, \quad \sum_{n-p+2}^n \alpha_i \geq p + q - 1, \quad \text{and} \quad \sum_{n-p+2}^n \beta_i = 2p - 2.$$

The line in this display that has $\Sigma_{n-k}^n \alpha_i \leq q + k$ and $\Sigma_{n-k-1}^n \alpha_i \geq q + k + 2$ requires either

$$\sum_{n-k-1}^n \beta_i \leq p + k + 1 \quad \text{or} \quad \sum_{n-k-1}^{n-1} \beta_i \geq p + k + 1$$

according to A_{k+2} in Lemma 1(a) to prevent $\mathscr{C}$ from being reducible. However, $\Sigma_{n-k}^n \alpha_i \leq q + k$ implies $\Sigma_{n-k}^{n-1} \alpha_i \leq q + k - 1$, or

$$\sum_{(n-1)-(k-1)}^{(n-1)} \alpha_i \leq q + (k - 1),$$

and it follows from the predecessor display for $n - 1$ in place of n that

$$\sum_{(n-1)-(k-1)-1}^{(n-1)} \beta_i \leq p + k, \quad \text{i.e.,} \quad \sum_{n-k-1}^{n-1} \beta_i \leq p + k.$$

Hence $\Sigma_{n-k-1}^{n-1} \beta_i \geq p + k + 1$ is precluded in the line indicated earlier for the n display, which is therefore correct as it stands.

It follows from $A_1 - A_p$ and irreducibility that for each $1 \leq n \leq N$ there is $0 \leq k \leq \min\{n - 1, p - 2\}$ such that

$$\sum_{n-k}^n \alpha_i \geq q + k + 1 \quad \text{and} \quad \sum_{n-k}^n \beta_i \leq p + k.$$

Beginning at N, proceed backwards through $\mathscr{C}$ as follows. Select $k_1 \geq 1$ for which

$$\sum_{k_1}^N \alpha_i \geq q + (N - k_1) + 1 \quad \text{and} \quad \sum_{k_1}^N \beta_i \leq p + (N - k_1).$$

If $k_1 > 1$, select $1 \le k_2 < k_1$ for which

$$\sum_{k_2}^{k_1-1} \alpha_i \ge q + (k_1 - 1 - k_2) + 1 \quad \text{and} \quad \sum_{k_2}^{k_1-1} \beta_i \le p + (k_1 - 1 - k_2),$$

and continue in the obvious way back to the beginning of $\mathscr{C}$. In each backwards step,

$$p(\prec\text{-excess}) - q(\sim\text{-excess}) \ge p(q) - q(p-1) = q,$$

so $pe(\prec) - qe(\sim) \ge q$ for $\mathscr{C}$. However, such a $\mathscr{C}$ is reducible: replace a $\prec^2$ segment by $\prec$ to get $\mathscr{C}'$ with $pe'(\prec) - qe'(\sim) \ge q - p \ge 0$. Therefore our initial supposition that $\mathscr{C}$ is not reducible is false. □

Proof of Lemma 3. Assume for definiteness that $\beta_1 \ge p + 1$, and suppose $\mathscr{C}$ is not reducible. For $n \le \min\{N, p\}$, a procedure like that in the preceding proof with Lemma 1(b) instead of Lemma 1(a) gives either

$\beta_n \ge p + 1$ and $\alpha_{n+1} \le q$; or

$\beta_n \le p$, $\beta_{n-1} + \beta_n \ge p + 2$, and $\alpha_n + \alpha_{n+1} \le q + 1$; or

$\vdots$

$$\sum_2^n \beta_i \le p + n - 2, \quad \sum_1^n \beta_i \ge p + n, \quad \text{and} \quad \sum_2^{n+1} \alpha_i \le q + n - 1.$$

At $n = p$, the first inequality in the final line implies $\beta_2 = \cdots = \beta_p = 2$. If $p < N$, a similar display with p lines for A_1 through A_p applies to each $n \in \{p+1, \ldots, N\}$. It follows that each $n \in \{1, \ldots, N\}$ has a $0 \le k \le \min\{n-1, p-1\}$ such that

$$\sum_{n-k}^n \beta_i \ge p + k + 1 \quad \text{and} \quad \sum_{n-k+1}^{n+1} \alpha_i \le q + k.$$

Beginning at N, proceed backwards: select $k_1 \ge 1$ for which

$$\sum_{k_1}^N \beta_i \ge p + (N - k_1) + 1 \quad \text{and} \quad \sum_{k_1+1}^{N+1} \alpha_i \le q + (N - k_1),$$

where $\alpha_{N+1} = \alpha_1$, then select $1 \le k_2 < k_1$ in a similar way if $k_1 > 1$, and so forth. In each step, $q(\sim\text{-excess}) - p(\prec\text{-excess}) \ge q(p) - p(q-1) = p$, so $qe(\sim) > pe(\prec)$ for $\mathscr{C}$. But then $\mathscr{C}$ is not forbidden, and it follows that a forbidden $\mathscr{C}$ is reducible when some $\beta_i \ge p + 1$. □

Proof of Lemma 4. Given the hypotheses of the lemma, we suppose that none of (a), (b), and (c) holds, and proceed to a contradiction.

Reducibility Lemmas

Suppose first that the α_j do not satisfy (a), and for definiteness take

$$\sum_{i=1}^{k+2} \alpha_i \geq q + k + 2.$$

Since $\sum_2^{k+2} \alpha_i \leq q + k$ and $\sum_1^{k+1} \beta_i \leq p + k$ by hypothesis, the presumed failure of (c) resulting from an application of A_{k+2} in Lemma 1(a) requires $\sum_1^{k+2} \beta_i \leq p + k + 1$. Continuing as in the proof of Lemma 2, for each $n \in \{k + 2, \ldots, N, N + 1, \ldots, N + k + 1\}$ we get either

$$\sum_{n-k}^{n} \alpha_i \leq q + k, \quad \sum_{n-k-1}^{n} \alpha_i \geq q + k + 2, \quad \sum_{n-k-1}^{n} \beta_i \leq p + k + 1; \quad \text{or}$$

$$\sum_{n-k-1}^{n} \alpha_i \leq q + k + 1, \quad \sum_{n-k-2}^{n} \alpha_i \geq q + k + 3,$$

$$\sum_{n-k-2}^{n} \beta_i \leq p + k + 2; \quad \text{or}$$

$$\vdots$$

$$\sum_{n-k-x}^{n} \alpha_i \leq q + k + x, \quad \sum_{n-k-x-1}^{n} \alpha_i \geq q + k + x + 2,$$

$$\sum_{n-k-x-1}^{n} \beta_i \leq p + k + x + 1,$$

where $x = \min\{(n - 2) - k, (p - 3) - k, (N - 2) - k\}$. As we proceed through larger values of n, the other possible inequality on the β_i sum that arises from the presumed failure of reducibility via Lemma 1(a) is precluded by the inequalities on the α_i sums of that case along with those obtained at $n - 1$.

It follows for each $n \in \{k + 2, \ldots, N + k + 1\}$ that there is a $k + 1 \leq y \leq \min\{n - 1, p - 2, N - 1\}$ such that

$$\sum_{n-y}^{n} \alpha_i \geq q + y + 1 \quad \text{and} \quad \sum_{n-y}^{n} \beta_i \leq p + y.$$

We therefore have $p(\prec \text{-excess}) - q(\sim \text{-excess}) \leq q$ in the part of $\mathscr{C}$ covered by these two inequalities. Although the backwards procedure used in the proof of Lemma 2 (begin at $N + k + 1$, get y for this ending point; take the next n as $N + k + 1 - y - 1$, get y for this n; ...) may not come out evenly by ending precisely at $k + 2$, we can continue around the picycle an arbitrarily large number of times and conclude that the average difference between $pe(\prec)$ and $qe(\sim)$ per revolution is at least q. Consequently, $\mathscr{C}$ must have $pe(\prec) - qe(\sim) \geq q$.

However, this would satisfy conclusion (b) of Lemma 4, so to maintain the supposition that none of (a), (b), and (c) holds, we need to suppose that the β_j do not satisfy (a). But then, given

$$\beta_i + \cdots + \beta_{i+k+1} \geq p + k + 2 \quad \text{for some } i,$$

a similar proof (see also the proof of Lemma 3) leads to the conclusion that $qe(\sim) > pe(\prec)$, which contradicts forbiddenness. As in the preceding paragraph, it may be necessary to cycle backwards around $\mathscr{C}$ a large number of times to conclude that $qe(\sim) > pe(\prec)$. We omit the details. □

8.6 BANISHING FORBIDDEN PICYCLES

We complete the sufficiency proof of Theorem 4 by showing that an interval order which satisfies $A[p,q]_1$ through $A[p,q]_p$ has no $[p,q]$-forbidden picycle. By Theorem 5, such an interval order is in $\mathscr{P}[p,q]$. As in the preceding section, the $[p,q]$ designation will often be omitted.

Assume henceforth that $(X, \prec)$ satisfies A_1–A_p. We shall suppose that $(X, \prec)$ has a forbidden picycle and obtain a contradiction.

By this supposition, $(X, \prec)$ has a minimum-length forbidden picycle, say $\mathscr{C}$ with index $(\alpha_1, \beta_1, \ldots, \alpha_N, \beta_N)$. Since $\mathscr{C}$ is not reducible, Lemma 1(i) says that $\alpha_i \geq 2$ and $\beta_i \geq 2$ for all i. If $p = 1$ (unitary class), then $\mathscr{C}$ has $e(\prec) \geq qe(\sim)$, and a contradiction to irreducibility follows immediately from the $(\sim \prec^{q+1}) \subseteq \prec$ part of A_1 since $\alpha_i - 1 \geq q$ for some i. *This completes the sufficiency proof of Theorem 2.*

Assume henceforth that $p \geq 2$. Then $q > p$, and therefore $\alpha_i \geq 3$ for some i. Assume that $\alpha_1 \geq 3$ for definiteness. Ensuing paragraphs consider possibilities for N versus p.

Suppose first that $N \geq p$. Then Lemmas 2 and 3, and induction with Lemma 4 if $p \geq 3$, give

$$\sum_{j=i}^{i+p-2} \alpha_j \leq p + q - 2 \quad \text{and} \quad \sum_{j=i}^{i+p-2} \beta_j \leq p + p - 2$$

for $i = 1, \ldots, N$, where as usual $\alpha_j = \alpha_{j-N}$ and $\beta_j = \beta_{j-N}$ for $j > N$. The inequality on the β_j implies that $\beta_i = 2$ for all i. Then, in view of Lemma 1(b) for A_p, irreducibility of $\mathscr{C}$ requires either $\sum_i^{i+p-1} \alpha_j \leq p + q - 1$ or $\sum_i^{i+p-2} \alpha_j \geq p + q - 1$ for each i. Since the latter inequality is false,

$$\sum_{j=i}^{i+p-1} \alpha_j \leq p + q - 1 \quad \text{for } i = 1, \ldots, N.$$

Therefore $\alpha_1 + \cdots + \alpha_N \leq N(p + q - 1)/p$. But then

$$pe(\prec) \leq p[N(p + q - 1)/p - N] = N(q - 1) < Nq = ne(\sim),$$

Banishing Forbidden Picycles

which contradicts forbiddenness. Hence $N \geq p$ yields a contradiction to our supposition that $(X, \prec)$ has a forbidden picycle.

Assume henceforth that $N < p$. By Lemmas 2 and 3 if $N = 1$, and by Lemmas 2 and 3 and induction with Lemma 4 if $N \geq 2$, we get

$$\sum_{i=1}^{N} \alpha_i \leq q + N - 1 \quad \text{and} \quad \sum_{i=1}^{N} \beta_i \leq p + N - 1. \tag{1}$$

Let

$$\Delta = pe(\prec) - qe(\sim).$$

Since $\mathscr{C}$ is irreducible, $p > \Delta \geq 0$. Let $\mathscr{C}_0$ be the picycle obtained from $\mathscr{C}$ by shortening the initial chain in $\mathscr{C}$ from length α_1 to $\alpha_1 - 1$. $\mathscr{C}_0$ has index $(\alpha_1 - 1, \beta_1, \ldots, \alpha_N, \beta_N)$ and is *not* forbidden since

$$p(\prec \text{-excess of } \mathscr{C}_0) - q(\sim \text{-excess of } \mathscr{C}_0) = \Delta - p < 0.$$

We now form forbidden picycles $\mathscr{C}^{(s)}$ with excesses $e^{(s)}(\prec)$ and $e^{(s)}(\sim)$ for $s = 1, 2, \ldots$, by repetitions of $\mathscr{C}$, mixed with repetitions of $\mathscr{C}_0$ when $\Delta > 0$. In what follows, $\mathscr{C}\mathscr{C}$ is $\mathscr{C}$ followed by a copy of itself; it has index $(\alpha_1, \beta_1, \ldots, \alpha_N, \beta_N, \alpha_1, \beta_1, \ldots, \alpha_N, \beta_N)$. Likewise, $\mathscr{C}\mathscr{C}_0$ denotes $\mathscr{C}$ followed by $\mathscr{C}_0$. It has index $(\alpha_1, \ldots, \beta_N, \alpha_1 - 1, \beta_1, \ldots, \beta_N)$ and length $2\sum_{1}^{N}(\alpha_i + \beta_i) - 1$. Expressions $\mathscr{C}\mathscr{C}\mathscr{C}_0, \mathscr{C}\mathscr{C}\mathscr{C}, \mathscr{C}\mathscr{C}_0\mathscr{C}, \ldots$ are defined similarly.

Let $\mathscr{C}^{(1)} = \mathscr{C}$, and for each $s \geq 1$ take

$$\mathscr{C}^{(s+1)} = \begin{cases} \mathscr{C}^{(s)}\mathscr{C} & \text{if } p > pe^{(s)}(\prec) - qe^{(s)}(\sim) + \Delta, \\ \mathscr{C}^{(s)}\mathscr{C}_0 & \text{if } pe^{(s)}(\prec) - qe^{(s)}(\sim) + \Delta \geq p, \end{cases}$$

so that $p > pe^{(s)}(\prec) - qe^{(s)}(\sim) \geq 0$ for all s. The index of $\mathscr{C}^{(s)}$ will be written as

$$\left(\alpha_1^{(s)}, \beta_1^{(s)}, \ldots, \alpha_{sN}^{(s)}, \beta_{sN}^{(s)}\right),$$

and subscripts on α and β will be taken modulo sN when they exceed sN. It may be noted that each $\mathscr{C}^{(s)}$ is forbidden, and no $\mathscr{C}^{(s)}$ is reducible by shortening a chain. Hence the type of reducibility specified in Lemma 4(b) never applies to $\mathscr{C}^{(s)}$.

We consider $s = 2, 3$ first, then generalize to larger s. Our aim is to prove that the "shortest forbidden picycle" supposition for $\mathscr{C}$ along with $1 \leq N < p$ forces the contradiction that p is infinite.

Let $s = 2$. Then $\mathscr{C}^{(2)}$ is $\mathscr{C}\mathscr{C}$ or $\mathscr{C}\mathscr{C}_0$. In either case, (1) implies that

$$\sum_{j=i}^{i+N-1} \alpha_j^{(2)} \leq q + N - 1 \quad \text{and} \quad \sum_{j=1}^{i+N-1} \beta_j^{(2)} \leq p + N - 1 \tag{2}$$

for $i = 1, \ldots, 2N$. We eliminate the case of $N = p - 1$ before considering other cases.

Suppose $N = p - 1$. Then, by (1), $\beta_i = 2$ for all i, so $\beta_i^{(2)} = 2$ for all i. In addition, $\sum_1^N \alpha_i = \sum_1^{p-1} \alpha_i \leq p + q - 2$, and therefore

$$\sum_{j=i}^{i+N-1} \alpha_j^{(2)} \leq p + q - 2 \quad \text{for } i = 1, \ldots, 2N.$$

If $\sum_i^{i+N} \alpha_j^{(2)} \leq p + q - 1$ for $i = 1, \ldots, 2N$ then, as in the paragraph preceding (1), we contradict forbiddenness:

$$pe^{(2)}(\prec) \leq p[2N(p + q - 1)/p - 2N] = 2N(q - 1) < 2Nq = qe^{(2)}(\sim).$$

Therefore $\sum_i^{i+N} \alpha_j^{(2)} \geq p + q$ for some i. Then, according to Lemma 1 with $n = p$, $\mathscr{C}^{(2)}$ is reducible to $\mathscr{C}'$ with

$$h(\mathscr{C}') = h(\mathscr{C}^{(2)}) - [p + q + 2(p - 1)] < h(\mathscr{C}),$$

where the inequality follows from the fact that

$$h(\mathscr{C}^{(2)}) - h(\mathscr{C}) \leq h(\mathscr{C}) \leq (p + q - 2) + 2(p - 1).$$

But then $\mathscr{C}'$ is a forbidden picycle that is shorter than $\mathscr{C}$, contradicting our minimum-length supposition. Therefore $N = p - 1$ is impossible.

Assume henceforth that $N \leq p - 2$. Given (2), apply Lemma 4 to $\mathscr{C}^{(2)}$ with $2N$ in place of N for this application. Since $N - 1 \leq p - 3$ and $(N - 1) + 2 \leq 2N$, and since (2) holds, the hypotheses of Lemma 4 for $\mathscr{C}^{(2)}$ hold for $k = N - 1$. Since $p \geq pe^{(2)}(\prec) - qe^{(2)}(\sim)$ by construction, conclusion (b) of Lemma 4 is false. Therefore either (a) or (c) holds with $k = N - 1$. That is, either

$$\sum_{j=i}^{i+N} \alpha_j^{(2)} \leq q + N \quad \text{and} \quad \sum_{j=i}^{i+N} \beta_j^{(2)} \leq p + N \quad \text{for } i = 1, \ldots, 2N, \quad (3)$$

or else $\mathscr{C}^{(2)}$ is reducible to $\mathscr{C}'$ with $h(\mathscr{C}') = h(\mathscr{C}^{(2)}) - [p + q + 2(n - 1)]$ for some $n \geq k + 2 = N + 1$. Suppose the latter alternative holds. Then

$$h(\mathscr{C}') \leq h(\mathscr{C}^{(2)}) - (p + q + 2N).$$

By (1), $h(\mathscr{C}) \leq (q + N - 1) + (p + N - 1) = p + q + 2N - 2$. Since $h(\mathscr{C}^{(2)}) - h(\mathscr{C})$ is either $h(\mathscr{C})$ or $h(\mathscr{C}) - 1$, it follows that $h(\mathscr{C}^{(2)}) - h(\mathscr{C}) \leq p + q + 2N - 2$, that is, that

$$h(\mathscr{C}^{(2)}) - (p + q + 2N) \leq h(\mathscr{C}) - 2.$$

Therefore $h(\mathscr{C}') \leq h(\mathscr{C}) - 2$. But this contradicts minimality for $\mathscr{C}$.

Therefore (3) holds. By repeating the use of Lemma 4 applied to $\mathscr{C}^{(2)}$ for increasing values of $k \geq N - 1$, we conclude that, with $K = \min\{p - 3, 2N - 2\}$,

$$\sum_{j=i}^{i+K+1} \alpha_j^{(2)} \leq q + K + 1 \quad \text{and} \quad \sum_{j=i}^{i+K+1} \beta_j^{(2)} \leq p + K + 1$$

for $i = 1, \ldots, 2N$. Suppose here that $K = p - 3$, so that

$$\sum_{j=i}^{i+p-2} \alpha_j^{(2)} \leq p + q - 2 \quad \text{and} \quad \sum_{j=i}^{i+p-2} \beta_j^{(2)} \leq p + p - 2$$

for $i \leq 2N$. Then $\beta_i^{(2)} = 2$ for all i and, as in the paragraph following (2), we obtain a contradiction: either $\sum_i^{i+p-1} \alpha_j^{(2)} \leq p + q - 1$ for all i, which contradicts forbiddenness, or else $\sum_i^{i+p-1} \alpha_j^{(2)} \geq p + q$ for some i, in which case Lemma 1 with $n = p$ shows that $\mathscr{C}^{(2)}$ is reducible to $\mathscr{C}'$ with $h(\mathscr{C}') < h(\mathscr{C})$. Therefore $K = 2N - 2 < p - 3$, so that $2N \leq p - 2$ with

$$\sum_1^{2N} \alpha_i^{(2)} \leq q + 2N - 1 \quad \text{and} \quad \sum_1^{2N} \beta_i^{(2)} \leq p + 2N - 1. \tag{4}$$

Assume henceforth that $2N \leq p - 2$ along with (4), and consider $\mathscr{C}^{(3)}$ for $s = 3$. According to the definition of $\mathscr{C}^{(3)}$, each $\sum_i^{i+2N-1} \beta_j^{(3)}$ includes each β_i ($i = 1, \ldots, N$) exactly twice so that these sums are equal: (4) gives

$$\sum_i^{i+2N-1} \beta_j^{(3)} \leq p + 2N - 1 \quad \text{for } i = 1, \ldots, 3N. \tag{5}$$

In a similar manner, if $\mathscr{C}^{(3)}$ is anything other than $\mathscr{C}\mathscr{C}_0\mathscr{C}$, then the first inequality in (4) implies that

$$\sum_i^{i+2N-1} \alpha_j^{(3)} \leq q + 2N - 1 \quad \text{for } i = 1, \ldots, 3N. \tag{6}$$

When this is true, an analysis like that in the preceding two paragraphs (applied to $\mathscr{C}^{(3)}$ with $3N$ in place of N for Lemma 4) implies that $3N \leq p - 2$ with

$$\sum_1^{3N} \alpha_i^{(3)} \leq q + 3N - 1 \quad \text{and} \quad \sum_1^{3N} \beta_i^{(3)} \leq p + 3N - 1. \tag{7}$$

The only way for (6) to fail is to have $\mathscr{C}^{(3)} = \mathscr{C}\mathscr{C}_0\mathscr{C}$ and $\sum_1^{2N} \alpha_i^{(2)} = q + 2N - 1$. We consider this further in the next paragraph.

Suppose $\mathscr{C}^{(3)} = \mathscr{C}\mathscr{C}_0\mathscr{C}$ with $\sum_1^{2N} \alpha_i^{(2)} = q + 2N - 1$. Then instead of (6) we get

$$\sum_i^{i+2N-1} \alpha_j^{(3)} = \begin{cases} q + 2N - 1 & \text{if } i \leq N + 1 \text{ or } i \geq 2N + 2, \\ q + 2N & \text{if } N + 2 \leq i \leq 2N + 1. \end{cases}$$

Since $\alpha_i^{(3)} \geq 2$ for all i, it is true that

$$\sum_i^{i+2N-2} \alpha_j^{(3)} \leq q + 2N - 2 \quad \text{for } i = 1, \ldots, 3N.$$

Therefore, taking account of (5) with $\beta_i^{(3)} \geq 2$, the hypotheses of Lemma 4 for $\mathscr{C}^{(3)}$ hold at $k = 2N - 2$. Since (b) of Lemma 4 is false by construction, it follows for $k = 2N - 2$ that either (a) holds, that is, either (5) and (6) hold, or else (c) holds with $\mathscr{C}^{(3)}$ reducible to $\mathscr{C}'$ via Lemma 1 for some $n \in \{k + 2, \ldots, \min\{p, 3N\}\}$ with $h(\mathscr{C}') = h(\mathscr{C}^{(3)}) - [p + q + 2(n - 1)]$. Suppose (c) holds. Then it could hold for $n = k + 2 = 2N$ only if $\sum_i^{i+2N-1} \beta_j^{(3)} \geq p + 2N$ for some i [to apply A_{2N} in Lemma 1], and this is false by (5). Therefore (c) requires $n \geq 2N + 1$, in which case

$$h(\mathscr{C}') \leq h(\mathscr{C}^{(3)}) - (p + q + 4N) = 3h(\mathscr{C}) - 1 - (p + q + 4N),$$

so that $h(\mathscr{C}') - h(\mathscr{C}) \leq 2h(\mathscr{C}) - 1 - (p + q + 4N)$. In fact, $\mathscr{C}^{(2)} = \mathscr{C}\mathscr{C}_0$ and the presumed $\sum_1^{2N}\alpha_i^{(2)} = q + 2N - 1$ give $2h(\mathscr{C}) = \sum_1^{2N}(\alpha_i^{(2)} + \beta_i^{(2)}) + 1 \leq p + q + 4N - 1$, so $2h(\mathscr{C}) - 1 - (p + q + 4N) \leq -2$. Therefore $h(\mathscr{C}') < h(\mathscr{C})$, contrary to minimality for $\mathscr{C}$, and so we conclude that (5) and (6) hold as stated.

Having arrived at (7) with $3N \leq p - 2$, we proceed by induction. The induction hypothesis is $sN \leq p - 2$ and

$$\sum_1^{sN} \alpha_i^{(s)} \leq q + sN - 1 \quad \text{and} \quad \sum_1^{sN} \beta_i^{(s)} \leq p + sN - 1 \quad (8)$$

for $s \geq 3$. We wish to show that $(s + 1)N \leq p - 2$ and that (8) holds with $s + 1$ in place of s. Given (8), the construction of $\mathscr{C}^{(2)}$ with repetitions of the original β_i gives

$$\sum_{j=i}^{i+sN-1} \beta_j^{(s+1)} \leq p + sN - 1 \quad \text{for } i = 1, \ldots, (s+1)N. \quad (9)$$

We also want

$$\sum_{j=i}^{i+sN-1} \alpha_j^{(s+1)} \leq q + sN - 1 \quad \text{for } i = 1, \ldots, (s+1)N. \quad (10)$$

Banishing Forbidden Picycles

This follows from (8) unless $\mathscr{C}^{(s+1)}$ has an internal $\mathscr{C}_0$ and ends with $\mathscr{C}$, as in $\mathscr{C}^{(s+1)} = \mathscr{C} \cdots \mathscr{C}\mathscr{C}_0 \cdots \mathscr{C}$, and the first inequality in (8) is an equality. Then the sum in (10) will equal $q + sN$ when i lies in a $\mathscr{C}_0$ block after the initial $\alpha_1 - 1$ in that block. Moreover, $h(\mathscr{C}^{(s+1)}) = h(\mathscr{C}^{(s)}) + h(\mathscr{C}) \le p + q + 2sN - 2$, the latter by (8).

Suppose (10) is violated in this way. Then the hypotheses of Lemma 4 for $\mathscr{C}^{(s+1)}$ hold at $k = sN - 2$, and we conclude from the lemma that either (9) and (10) hold or else $\mathscr{C}^{(s+1)}$ is reducible to $\mathscr{C}'$ via Lemma 1 with $h(\mathscr{C}') = h(\mathscr{C}^{(s+1)}) - [p + q + 2(n-1)]$ for some $n \ge k + 2 = sN$. However, this can hold at $n = sN$ only if $\sum_i^{i+sN-1} \beta_j^{(s+1)} \ge p + sN$ for some i (to apply A_{sN} in Lemma 1), and this is false by (9). Therefore a reduction from $\mathscr{C}^{(s+1)}$ to $\mathscr{C}'$ requires $n \ge sN + 1$, in which case

$$h(\mathscr{C}') \le h(\mathscr{C}^{(s+1)}) - (p + q + 2sN)$$
$$= h(\mathscr{C}^{(s)}) + h(\mathscr{C}) - (p + q + 2sN)$$
$$\le h(\mathscr{C}) - 2,$$

contrary to minimality for $\mathscr{C}$. Therefore (9) and (10) hold.

Given (9) and (10), apply Lemma 4 to $\mathscr{C}^{(s+1)}$, beginning at $k = sN - 1$. Since conclusion (b) never holds and conclusion (c) at any step contradicts the minimality of $\mathscr{C}$, we require (a) in all cases. If $\min\{p - 3, (s+1)N - 2\} = p - 3$, then a contradiction obtains as in the paragraph preceding (4). Hence $(s+1)N - 2 \le p - 4$, or $(s+1)N \le p - 2$, along with

$$\sum_1^{(s+1)N} \alpha_i^{(s+1)} \le q + (s+1)N - 1 \quad \text{and} \quad \sum_1^{(s+1)N} \beta_i^{(s+1)} \le p + (s+1)N - 1,$$

which is conclusion (a) for $k = (s+1)N - 2$. This verifies the desired induction conclusions.

It now follows that $p \ge sN$ for all integers s, which is absurd since $N \ge 1$. Therefore $(X, \prec)$ has no minimum-length forbidden picycle and hence no forbidden picycle.

9

Numbers of Lengths

In contrast to the bounded-length classes of the preceding chapter, we now consider classes of finite interval orders and interval graphs whose members have real representations that use at most n different interval lengths. It will be shown that, when $n \geq 2$, these classes cannot be characterized by finite sets of forbidden interval order restrictions or subgraphs. Efficient characterizations of our new classes for $n \geq 2$ do not presently exist. The later sections of the chapter discuss anomalies that arise in the finite-lengths context. The next chapter takes a closer look at the smallest interval orders whose representations require at least k different lengths for $k = 2, 3 \ldots$.

9.1 FINITE-LENGTHS CLASSES

For every positive integer n let

$\mathscr{P}_n = \{(X, \prec): (X, \prec)$ is a finite interval order some representation of which has $|\rho(X)| \leq n\}$,

$\mathscr{I}_n = \{(X, \sim): (X, \sim)$ is a finite interval graph some representation of which has $|\rho(X)| \leq n\}$.

The definition in Section 5.6 of $\rho^*(X, \prec)$ as the minimum number of lengths needed to represent a finite interval order $(X, \prec)$ gives

$$\mathscr{P}_n = \{(X, \prec): \rho^*(X, \prec) \leq n\}.$$

In addition, for any finite interval graph $(X, \sim)$, let

$$\rho^*(X, \sim) = \min\{\rho^*(X, \prec): (X, \prec) \text{ is an interval order that agrees with } (X, \sim)\}.$$

Then

$$\mathscr{I}_n = \{(X, \sim): \rho^*(X, \sim) \leq n\}.$$

Finite-Lengths Classes

It is easily checked that $\mathscr{P}_1, \mathscr{P}_2, \ldots$ and $\mathscr{I}_1, \mathscr{I}_2, \ldots$ are strictly increasing sequences. Indeed, Theorem 5.18 says that every depth-n interval order on $3n - 2$ points has $\rho^*(X, \prec) = n$, so such an order is in $\mathscr{P}_n$ but not in $\mathscr{P}_k$ for $k < n$. Moreover, you can readily verify that the symmetric-complement graph of such an order has $\rho^*(X, \sim) = n$, so $\mathscr{I}_1 \subset \mathscr{I}_2 \subset \ldots$.

Clearly, $\mathscr{P}_1 = \mathscr{P}[1,1]$, the class of finite semiorders, and $\mathscr{I}_1 = \mathscr{I}[1,1]$, the class of finite indifference graphs. However, for $n \geq 2$, $\mathscr{P}_n$ and $\mathscr{I}_n$ are very different from any of the bounded-length classes $\mathscr{P}[p,q]$ and $\mathscr{I}[p,q]$ of the preceding chapter as will be evident from the examples and theorems given below. For example, unlike the case for rational q/p in Corollaries 8.1 and 8.2, $\mathscr{P}_n$ and $\mathscr{I}_n$ for $n \geq 2$ cannot be characterized by finite sets of forbidden interval order restrictions and forbidden subgraphs. This will be proved in Sections 9.2 and 9.3. Sections 9.4 and 9.5 then look at paradoxical aspects of finite lengths. In particular, we shall see that $\rho^*(X, \prec)$ can be arbitrarily large for depth-2 interval orders and that some interval orders in $\mathscr{P}_2$ do not have continuous ranges for the admissible longer length when the shorter length is fixed at 1.

We conclude this introduction with three examples that illustrate important facets of the finite-lengths setting.

Example 1. Unlike magnitudes and the agreement between $\mathscr{I}[p,q]$ and $\mathscr{P}[p,q]$ in Theorem 8.1, it need not be true that $\rho^*(X, \sim) = \rho^*(X, \prec)$ for every interval order $(X, \prec)$ that agrees with interval graph $(X, \sim)$. The interval graph pictured at the top of Fig. 9.1 has $\rho^*(X, \sim) = 2$, but its agreeing interval order at the bottom of the figure has $\rho^*(X, \prec) = 3$ since depth$(X, \prec) = 3$. Note also that the minimal buried subgraph in $(X, \sim)$ is $\{d, e, g, h, i\}$ with K-set $\{c\}$ and, within this subgraph, $\{e, g\}$ is a buried

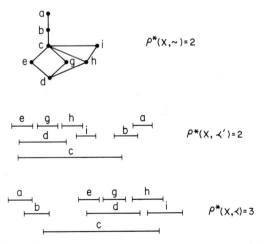

Figure 9.1 Interval graph and two agreeing interval orders.

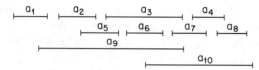

Figure 9.2 A depth-2 interval order with $\rho^* = 3$.

subgraph with K-set $\{d\}$. Consonant with Theorem 8.1, the agreeing interval orders are in $\mathcal{P}[1,q]$ for every $q > 2$, and $(X, \sim)$ is in $\mathcal{I}[1,q]$ for every $q > 2$.

Example 2. Figure 9.2 describes the smallest depth-2 interval order that is not in $\mathcal{P}_2$. Suppose only two lengths, say $\rho_1 < \rho_2$, are needed for the representation. Then a_2 and a_7 have ρ_1, while a_3 and a_9 have ρ_2. But this is impossible since the right end of a_9 must be strictly within ρ_1 units of the right end of a_3, but the left end of a_9 must be more than ρ_1 units to the left of the left end of a_3. Generalizations of the idea that drives this result are used in the later proofs of Theorems 1 and 5.

Example 3. As before, let $<^-$ and $<^+$ be the lexicographic left and right endpoint orders defined by $x <^- y$ if $x \prec^- y$ or $(x \sim^- y, x \prec^+ y)$, and $x <^+ y$ if $x \prec^+ y$ or $(x \sim^+ y, x \prec^- y)$. It is *not* true that every interval order has a minimum-lengths representation in which intervals' left endpoints are ordered by $<^-$ and their right endpoints are ordered by $<^+$. A case in point is shown in Fig. 9.3. This interval order is in $\mathcal{P}_2$ as suggested by the figure. Suppose for definiteness that the shorter length is 1. Then c forces d's right endpoint to exceed b's right endpoint by more than 1 and, since $a \sim d$, a's left endpoint must exceed b's left endpoint *if* a is a short interval. But this contradicts $a <^- b$ as given by $a \sim^- b$ and $a \prec^+ b$.

This violation of $<^-$ could only be avoided if a were given the longer length. However, there would then be no two-lengths representation, because, if there were, a would have to intersect y, and in fact $y \prec a$. To verify this, suppose there is a two-lengths representation with a having the longer length $\beta > 1$. By d and x respectively, $4 < \beta < 5$. Therefore a's right endpoint is within 2 units of b's left endpoint, and then a's left endpoint (with $a <^- b$) is more than two units to the left of b's left endpoint. However, since e's left endpoint is within 5 units of b's left endpoint, a and y must intersect.

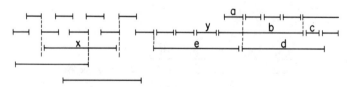

Figure 9.3 b begins before a, violating $a <^- b$.

9.2 TWO LENGTHS

Leibowitz (1978) analyzes interval graphs from the viewpoint of $\mathscr{I}_n$ and identifies three families of interval graphs in $\mathscr{I}_2$. They are interval graphs with subgraphs on all but one point that are indifference graphs, trees (connected, acyclic graphs) that are interval graphs, and threshold graphs (Chvátal and Hammer, 1973; Golumbic, 1980). However, the problems of characterizing $\mathscr{I}_2$ and $\mathscr{P}_2$ remain open.

One reason for this is revealed in the following theorems, which are extended to $n \geq 3$ in the next section.

Theorem 1. *Suppose $\mathscr{P}_0$ is a set of interval orders such that, for every interval order $(X, \prec)$, $(X, \prec)$ is in $\mathscr{P}_2$ if and only if no restriction of $(X, \prec)$ is isomorphic to an order in $\mathscr{P}_0$. Then $\mathscr{P}_0$ is infinite.*

Theorem 2. *Suppose $\mathscr{I}_0$ is a set of interval graphs such that, for every interval graph $(X, \sim)$, $(X, \sim)$ is in $\mathscr{I}_2$ if and only if no subgraph of $(X, \sim)$ is isomorphic to a graph in $\mathscr{I}_0$. Then $\mathscr{I}_0$ is infinite.*

Proof of Theorem 1. We construct interval orders of arbitrary large finite cardinality that are not in $\mathscr{P}_2$ but have every proper restriction in $\mathscr{P}_2$. Then $\mathscr{P}_0$ as described in the theorem must be infinite.

Let $(X, \prec)$ be the interval order on the $2r + 7$ points in $X = \{x, y, c, d, a_0, a_1, \ldots, a_{r+1}, b_1, \ldots, b_{r+1}\}$ whose representation is shown in Fig. 9.4. It has $c \prec x \prec b_1 \prec \cdots \prec b_{r+1}, a_0 \prec a_1 \prec a_3, a_2 \prec a_4, a_{r-2} \prec a_r, a_{r+1} \prec b_1, a_{r-1} \prec y$, and so forth. Assume that $r \geq 3$. If $(X, \prec) \in \mathscr{P}_2$, then $a_1, \ldots, a_{r+1}, b_1, \ldots,$ and b_r must have the shorter length. But then x and y could not have the same length since, with ρ_1 as the shorter length, y's left endpoint is within $r\rho_1$ units of x's left endpoint while y's right endpoint is more than $r\rho_1$ units greater than x's right endpoint. Therefore $(X, \prec)$ is not in $\mathscr{P}_2$.

To prove that every proper restriction of $(X, \prec)$ is in $\mathscr{P}_2$, we need only consider restrictions that remove one point from X. These are outlined as

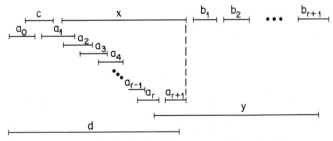

Figure 9.4 Not in $\mathscr{P}_2$; all proper restrictions in $\mathscr{P}_2$.

follows.

1. Remove x. All points except d and y get the shorter length; d and y are assigned the longer length.
2. Remove y. All but d and x get the shorter length.
3. Remove d. All but a_1, x, and y are short. The longer length for a_1 allows x and y to have the same longer length.
4. Remove c. All but d, x, and y are short. Begin a_1 to the right of the left endpoint of x so that x and y can have the same longer length.
5. Remove a_0. Make a_1, x, y, and d long.
6. Remove one of a_1 through a_r. This leaves a $\prec$ gap in the $a_1, \ldots, a_r, y$ series; x, y, and d can be given the same longer length.
7. Remove a_{r+1}. Assign unit lengths to all but x, y, d, and a_r, which get length N:

$$F(c) = [-1, 0]$$
$$F(a_0) = [-1-\lambda, -\lambda]$$
$$F(a_i) = [i-1, i] \quad 1 \le i \le r-1$$
$$F(b_i) = [r-2+i+N+i\lambda, r-1+i+N+i\lambda] \quad 1 \le i \le r+1$$
$$F(x) = [\lambda, N+\lambda]$$
$$F(y) = [N-\lambda, 2N-\lambda]$$
$$F(d) = [-\lambda, N-\lambda]$$
$$F(a_r) = [r-1, r-1+N].$$

This is easily seen to suffice when $\lambda > 0$ is small and N is large.

8. Remove one of the b_i. Renumber the r other b_i as $b_1, \ldots, b_r$. Use the preceding representation for c, a_0, and a_1 through a_{r-1} along with

$$F(a_r) = [r-1, r]$$
$$F(a_{r+1}) = [r+1, r+2]$$
$$F(b_i) = [N-1+i+(i+1)\lambda, N+i+(i+1)\lambda] \quad 1 \le i \le r$$
$$F(x) = [\lambda, N+\lambda]$$
$$F(y) = [r, r+N]$$
$$F(d) = [r+1-N, r+1].$$

Again, small λ and large N give the desired result. (For y to intersect b_r, we need $r + N \geq N - 1 + r + (r + 1)\lambda$, or $\lambda \leq 1/(r + 1)$.)

Cases 1 through 8 account for all points in x. Hence, for each $r \geq 3$, $(X, \prec)$ is not in $\mathscr{P}_2$ but each proper restriction is in $\mathscr{P}_2$. □

Proof of Theorem 2. Given $r \geq 3$, let $(X, \sim)$ be the symmetric-complement graph of the interval order in Fig. 9.4. Then $\{b_1, \ldots, b_{r+1}\}$ with K-set $\{y\}$ is the only maximal buried subgraph of $(X, \sim)$. By Theorem 3.11, interval orders that agree with $(X, \sim)$ are unique up to duality and permutations on the order of b_1 through b_{r+1}. Hence, by the preceding proof, $(X, \sim)$ has no two-lengths representation, so it is not in $\mathscr{I}_2$. However, every proper subgraph is in $\mathscr{I}_2$. Since r can be arbitrarily large, it follows that $\mathscr{I}_0$ for Theorem 2 must be infinite. □

We shall return to two-lengths representations in the final section of the chapter where we consider sets of admissible longer lengths when the shorter length is fixed at 1.

9.3 MANY LENGTHS

The following inductive extension theorems show that Theorems 1 and 2 hold for $\mathscr{P}_n$ and $\mathscr{I}_n$ ($n \geq 3$) as well as for $\mathscr{P}_2$ and $\mathscr{I}_2$.

Theorem 3. *Suppose $n \geq 3$ and there is a finite set $\mathscr{P}_0$ of interval orders such that an interval order $(X, \prec)$ is in $\mathscr{P}_n$ if and only if no restriction of $(X, \prec)$ is isomorphic to an order in $\mathscr{P}_0$. Then the same thing is true for $\mathscr{P}_{n-1}$ in place of $\mathscr{P}_n$.*

Theorem 4. *Suppose $n \geq 1$ and there is no finite set $\mathscr{I}_0$ of interval graphs such that an interval graph $(X, \sim)$ is in $\mathscr{I}_n$ if and only if no subgraph of $(X, \sim)$ is isomorphic to a graph in $\mathscr{I}_0$. Then the same thing is true for $\mathscr{I}_{n+1}$ in place of $\mathscr{I}_n$.*

Proof of Theorem 3. Given the hypotheses of Theorem 3, at least $n + 1$ lengths are needed to represent each order in $\mathscr{P}_0$. Let $\mathscr{P}_0^*$ consist of $\mathscr{P}_0$ and all restrictions of orders in $\mathscr{P}_0$ whose representations require at least n lengths. Clearly, $\mathscr{P}_0^*$ is finite. We claim that if interval order $(X, \prec)$ requires more than $n - 1$ lengths for every representation, then a restriction of $(X, \prec)$ is isomorphic to an order in $\mathscr{P}_0^*$. The conclusion of the theorem then follows from this claim.

To substantiate the claim, suppose n or more lengths are needed to represent interval order $(X, \prec)$. If more than n lengths are needed, $\mathscr{P}_0$ will serve under the hypotheses of the theorem. Assume henceforth that $(X, \prec)$ is in $\mathscr{P}_n$ but not $\mathscr{P}_{n-1}$.

Let $\{a,b,c\}$ be a three-element set disjoint from X, set Y equal to $X \cup \{a,b,c\}$, and define $\prec^*$ on Y by

$$\prec^* \;=\; \prec \;\cup\; \{(a,x)\colon x \in \{c\} \cup X\} \;\cup\; \{(x,c)\colon x \in \{a\} \cup X\}.$$

Then $a \prec^* X \prec^* c$, and b is universal for $(Y, \sim^*)$. At least $n+1$ lengths (but no more) are needed to represent $(Y, \prec^*)$ since b requires a longer length than every $x \in X$. Therefore an order $\prec'$ in $\mathscr{P}_0$ is isomorphic to a restriction $(Y', \prec^*)$ of $(Y, \prec^*)$. This restriction must contain b, since otherwise it would be representable by fewer than $n+1$ lengths. Let X' be the subset of Y' that is disjoint from $\{a,b,c\}$. Then $(X', \prec^*) = (X', \prec)$, and $(X', \prec)$ is in $\mathscr{P}_n \setminus \mathscr{P}_{n-1}$ and is isomorphic to a restriction $\prec''$ of $\prec'$. By definition, $\prec''$ is in $\mathscr{P}_0^*$. Hence a restriction of $(X, \prec)$ is isomorphic to an order in $\mathscr{P}_0^*$. □

Proof of Theorem 4. Suppose the hypotheses of Theorem 4 hold for n. Then for every positive integer k there is an interval graph on more than k points that is not in $\mathscr{I}_n$ but which has every proper subgraph in $\mathscr{I}_n$. Let G be such a graph for n with $N > k$ points. It is easily seen that G is in $\mathscr{I}_{n+1}$. Let G' consist of G and two disjoint copies of G, plus one more point that is universal for the whole. Whether or not G is connected, G' can be represented with $n+2$ lengths, but no fewer: in any $(n+1)$-lengths representation of G' without the universal point, the same longest length must be used in at least three components of G' (without the universal); when the universal is restored, one of the longest prior intervals is properly included in the universal's interval.

Let G'' be a minimal subgraph of G' that is in $\mathscr{I}_{n+2} \setminus \mathscr{I}_{n+1}$ so that every proper subgraph of G'' is in $\mathscr{I}_{n+1}$. Then G'' contains the universal point and at least one of the three copies of every point in G since otherwise G'' without the universal would be in $\mathscr{I}_n$. It follows that G'' has at least $\frac{1}{3}(3N) + 1 = N + 1$ points.

The conclusion of Theorem 4 follows since, for every k, there is an interval graph on $N' > k$ points that is not in $\mathscr{I}_{n+1}$ but has every proper subgraph in $\mathscr{I}_{n+1}$. □

9.4 DEPTH-2 INTERVAL ORDERS

Let $\mathscr{D}_2$ denote the class of all depth-2 finite interval orders. According to Sections 5.1 and 5.6, for every $(X, \prec)$ in $\mathscr{D}_2$ we have $x \subset_0 y$ for some x and y but never $x \subset_0 y \subset_0 z$, and $(X, \prec)$ is in $\mathscr{P}_n$ for some $n \geq 2$. Example 2 shows that there are $(X, \prec)$ in $\mathscr{D}_2$ that are in $\mathscr{P}_3$ but not $\mathscr{P}_2$. We shall now prove that there are depth-2 interval orders whose representations require arbitrarily large numbers of lengths. Additional results and an open problem conclude the section.

Depth-2 Interval Orders

Theorem 5. *For every positive integer n there is an interval order in $\mathcal{D}_2$ that is not in $\mathcal{P}_n$.*

Proof. Figure 9.5 pictures the left-hand end of an interval order in $\mathcal{D}_2$ that is constructed by the following iterative scheme. We begin with a and the x_i: by the unlabeled interval on the far left, each x_i begins after a begins and must intersect a. The x_i end in a staggered arrangement, as shown by the intervals labeled "b" and the others nearby. Each b now operates like the original a: for the topmost b, y_1 through y_{n+1} mimic x_1 through x_{n+1} and terminate near the c intervals shown on the upper right, which in the next step act like the b intervals in the preceding step. (Just as x_{n+1} has no b, the last y_i under each b has no c.) A new set of $n + 1$ intervals which mimic x_1 through x_{n+1} begin within each c, as suggested by z_1 through z_{n+1}, which are pictured as one interval.

All $n(n + 1)$ intervals of type y that begin within the n b's extend beyond the right endpoint of x_{n+1}. The c intervals for the second b are to the right of the c intervals for the first b, the c intervals for the third b are to the right of those for the second b, and so forth. After all n^2 c's have been positioned, the intervals like z_1 through z_{n+1} that begin within the c's are extended to the right of the final interval y_* under the last b, just as the y_i under all b extend beyond the end of x_{n+1}.

We continue in the obvious way. The next step has n^3 staggered d intervals after the ends of the z intervals, and $n + 1$ new intervals begin within each d and extend to the right. Counting a and the x_i as step 1, b's and the y_i as step 2, c's and the z_i as step 3, ..., the construction continues through n steps. The intervals that extend to the right for the nth step, of which there are $n + 1$ for each of the n^{n-1} "short" intervals (like $a, b, c, \ldots$) in the nth step, terminate in

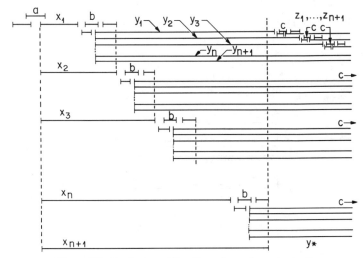

Figure 9.5 Beginning of iteration scheme for order in $\mathcal{D}_2$.

the manner shown in Fig. 9.6 with the addition of two intervals for each of the first n in each batch of $n + 1$.

Because of the staggered nature of the construction, there are no three points in the constructed interval order that form a C_0 chain, and hence the interval order is in $\mathcal{D}_2$. For notational convenience, let a_i stand for a generic "short" interval in step i: a is the only a_1; the n b's are the a_2's; the n^2 c's are the a_3's; .... The n^{n-1} a_n's for step n lie to the left of the vertical line in Fig. 9.6.

We claim that at least $n + 1$ different lengths are needed in every representation of the constructed interval order. To the contrary, suppose some representation requires only n lengths, say $\rho_1 < \rho_2 < \cdots < \rho_n$. Then, regardless of the lengths assigned to the little intervals to the right of the vertical line in Fig. 9.6, we must have

$$\rho(a_n) \geq \rho_2 \quad \text{for every } a_n.$$

For, if $\rho(a_n) = \rho_1$ for some a_n, then successively longer lengths must be used for the $n + 1$ intervals in its batch that cross the vertical of Fig. 9.6 (their left endpoints are all within ρ_1 units of each other), and this contradicts the supposition that only n lengths are needed.

Given $\rho(a_n) \geq \rho_2$ for every a_n, it then follows that $\rho(a_{n-1}) \geq \rho_3$ for every a_{n-1}, for if $\rho(a_{n-1}) \leq \rho_2$ for some a_{n-1} then the $n + 1$ intervals that begin within this a_{n-1} and extend to the right would all have different lengths since their left endpoints are within ρ_2 units of each other but their successive right endpoints are more than ρ_2 units apart. Continuing backwards, we get $\rho(a_2) \geq \rho_n$ for every a_2, that is, for every b. Hence

$$\rho(b) = \rho_n \quad \text{for every } b.$$

But this forces $\rho(a) > \rho_n$, since otherwise we must have $\rho(x_1) < \rho(x_2) < \cdots < \rho(x_{n+1})$.

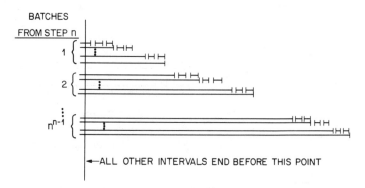

Figure 9.6 End of iteration scheme for order in $\mathcal{D}_2$.

Discontinuities and Admissible Lengths

Since it is impossible to represent the interval order using only n lengths, it is not in $\mathscr{P}_n$. □

Theorem 5 extends trivially to the class $\mathscr{D}_k$ of all depth-k finite interval orders for $k \geq 3$ by the insertion of additional intervals in the preceding construction. Moreover, similar results hold for interval graphs. Note in particular that the only buried subgraphs in the symmetric-complement graph of the constructed interval order are the two-point sets at the right ends of the long intervals in Fig. 9.6. It follows from Theorem 3.11 and the preceding proof that for every positive integer n there is an interval graph that is not in $\mathscr{I}_n$ and never has $x \subset_0 y \subset_0 z$ in an agreeing interval order.

In Example 2 it was claimed that the smallest interval order in $\mathscr{D}_2$ but not $\mathscr{P}_2$ has 10 points. Verification of this will not be given here. The question of the minimum cardinalities of interval orders in $\mathscr{D}_2$ that are not in $\mathscr{P}_n$ for $n = 3, 4, \ldots$ is wide open.

9.5 DISCONTINUITIES AND ADMISSIBLE LENGTHS

We conclude our discussion of numbers of lengths in the present chapter by examining two-lengths representations in greater detail. Our focus is on the set of admissible longer lengths in representations of interval orders in $\mathscr{P}_2 \setminus \mathscr{P}_1$ when only two lengths are used and the shorter length is fixed at 1. For each $(X, \prec) \in \mathscr{P}_2 \setminus \mathscr{P}_1$, let

$$\theta(X, \prec) = \{\alpha > 1 \colon \rho(X) = \{1, \alpha\} \text{ for some representation } (f, \rho) \text{ of } (X, \prec)\}.$$

Our first result for the longer-length function θ shows that for each integer $k \geq 2$ there is an interval order in $\mathscr{P}_2 \setminus \mathscr{P}_1$ for which $\theta = (1, k)$. After proving this we shall consider situations for which θ is not a continuous interval.

Theorem 6. *For every integer $k \geq 2$ there is an $(X, \prec)$ in $\mathscr{P}_2 \setminus \mathscr{P}_1$ for which $\theta(X, \prec) = (1, k)$.*

Proof. We consider $k \in \{2, 3\}$ first and then give a general scheme for $k \geq 4$.

The upper diagram in Fig. 9.7 is for $k = 2$. Intervals that must have length 1 in a two-lengths representation with lengths 1 and $\alpha > 1$ are marked by 1, and intervals that require the longer length are marked by α. The lowest α shows that $\alpha < 2$, and it is easily seen that every $1 < \alpha < 2$ is admissible so that $\theta(X, \prec) = (1, 2)$. For example, take $r_i = i$ for $i = 0, 1, 2$, $s_1 = \frac{1}{2}$ and $s_2 = \frac{3}{2}$. Given α in $(1, 2)$, all other intervals can be assigned length α and positioned to give a valid representation of $(X, \prec)$.

For $k = 3$, separate the two 1 intervals in the third row of the upper diagram: the first goes from r_0 to $r_0 + 1$; the second from $r_0 + 1 + \delta$ to

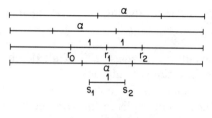

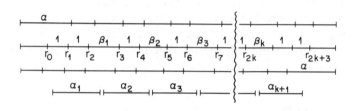

$\theta(X, \prec) = (1, k+3)$

Figure 9.7

$r_0 + 2 + \delta$. Since the three 1 intervals are still linked ($s_1 \leq r_0 + 1, r_0 + 1 + \delta \leq s_2$), the lowest α requires $\alpha < 3$. Values of α near 3 are obtained with $r_0 = 0$, $s_1 = \delta = 1$, $s_2 = 2$. Values of α near 1 are obtained with $0 < \delta < \alpha - 1$.

The lower diagram in Fig. 9.7 gives $\theta(X, \prec) = (1, k + 3)$ for $k \geq 1$. The second line has $2k + 5$ intervals placed end-to-end: there are two unmarked end intervals, $k + 3$ intervals marked 1, and k intervals marked β_1 through β_k. *The use of β rather than 1 or α to mark an interval indicates that the interval might be either a short or a long interval in a two-lengths representation.* The first line has a long interval at its left end and is otherwise a copy of the second line shifted slightly to the right. The third line ends with a long interval and is otherwise a copy of the second line shifted slightly to the left. The lowest line in the diagram identifies a chain of $k + 1$ intervals that must have the longer length α.

The left endpoint of α_1 exceeds r_0, the right endpoint of α_{k+1} is less than r_{2k+3}, and the successive 1's and β_i's in the second line successively intersect. Therefore

$$(k+1)\alpha < (k+3) + \sum_{i=1}^{k} \rho(\beta_i).$$

If $\rho(\beta_i) = \alpha$ for $i = 1, \ldots, k$, then $\alpha < k + 3$, so α can never be as great as $k + 3$.

Discontinuities and Admissible Lengths

We conclude the proof for the lower diagram by showing that all α in $(1, k + 3)$ are admissible. In the following construction, intervals in the first three lines are set end-to-end, as shown, and $\rho(\beta_i) = \alpha$ for all i.

Given $1 < \alpha < k + 3$, let

$$r_0 = 0$$

$$r_1 = 1$$

$$r_i = (i + 1)/2 + \alpha(i - 1)/2 \qquad i = 1, 3, 5, \ldots, 2k + 1$$

$$r_i = r_{i-1} + 1 = (i + 2)/2 + \alpha(i - 2)/2 \qquad i = 2, 4, \ldots, 2k + 2$$

$$r_{2k+3} = r_{2k+2} + 1 = k + 3 + \alpha k.$$

Let δ be positive and small. Position the internal endpoints in the first line at $r_i + \delta$ for $i = 1, 2, \ldots, 2k + 3$, and position the internal endpoints in the third line at $r_i - \delta$ for $i = 0, 1, \ldots, 2k + 2$. Given $0 < \mu < 1$, the α_i at the bottom of the diagram are positioned as follows:

$$\alpha_i \text{ goes from } i\mu + \alpha(i - 1) \text{ to } i\mu + \alpha i.$$

According to this, α_i is wholly to the left of α_{i+1}.

It is now routine to show that these placements satisfy the necessary inequalities for a valid representation when δ and μ are suitably chosen. In particular, the construction yields a valid representation if and only if

$$0 < \delta < (\alpha - 1)/2$$

$$0 < \mu < 1 - \delta$$

$$1 - (\alpha - 1 - \delta)/(k + 1) < \mu < 1 + (2 - \alpha)/(k + 1).$$

Comparisons of the bounds on μ show that there is a μ that satisfies these bounds if, given $\alpha > 1$,

$$\alpha < k + 3, \quad 0 < \delta < 1, \quad \text{and} \quad \delta \leq (\alpha - 1)/(k + 2).$$

Consequently, since $(\alpha - 1)/(k + 2) < 1$ exactly when $\alpha < k + 3$, the construction gives a valid representation for any δ in $(0, (\alpha - 1)/(k + 2)]$. □

We preface our theorems for discontinuous $\theta(X, \prec)$ by considering the seven intervals from a larger two-lengths interval order pictured in Fig. 9.8. As in the preceding proof, necessarily short intervals are marked by 1, necessarily long intervals are marked by α, and an interval that might be either short or long is marked by β. We refer to the latter as a *free interval*.

Suppose the free interval β, for x_4, is assigned the longer length α. Then, since

$$\rho(x_1) + \rho(x_2) > \rho(x_3) + \rho(x_4) + \rho(x_5),$$

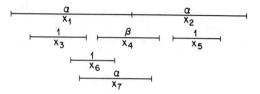

Figure 9.8 A free interval β.

$2\alpha > 1 + \alpha + 1$, or $\alpha > 2$. On the other hand, if β is assigned the shorter length 1 then, since

$$\rho(x_7) < \rho(x_4) + \rho(x_6),$$

$\alpha < 2$. Therefore $\alpha = 2$ is never admissible even though values of $\alpha < 2$ and values of $\alpha > 2$ may both be admissible.

The preceding example gives the essential key to the proofs of the following two theorems.

Theorem 7. *For every integer $k \geq 2$ there is an $(X, \prec)$ in $\mathscr{P}_2 \setminus \mathscr{P}_1$ for which*

$$\theta(X, \prec) = \left(2 - \frac{1}{k}, 2\right) \cup (k, \infty).$$

Theorem 8. *For every integer $k \geq 2$ there is an $(X, \prec)$ in $\mathscr{P}_2/\mathscr{P}_1$ for which*

$$\theta(X, \prec) = (k, 2k - 1) \cup (2k - 1, \infty).$$

Corollary 1. *For every $k \geq 1$ there is an $(X, \prec)$ in $\mathscr{P}_2 \setminus \mathscr{P}_1$ for which $\theta(X, \prec)$ is the union of k disjoint open intervals.*

The proof of the corollary is left to the reader. We conclude with the proofs of the theorems.

Proofs of Theorem 7. The proof for $k = 2$ is suggested by the example preceding the theorem. Figure 9.9 shows the interval order we shall use for $k \geq 3$. Necessarily short and long intervals are marked as before. The others $(x_0, x_{k+1}, c_i, e_i, f_i)$ could be either long or short.

The e_i are the key free intervals. If $\rho(e_i) = 1$, then $\alpha = \rho(g_i) < \rho(d_i) + \rho(e_i) = 2$, so $\alpha < 2$. If $\rho(e_i) = \alpha$, then $2 + \alpha = \rho(a_i) + \rho(a_{i+1}) + \rho(e_i) < \rho(x_i) + \rho(x_{i+1}) = 2\alpha$, so $2 < \alpha$. Hence a two-lengths representation requires either $\rho(e_i) = 1$ for all i or $\rho(e_i) = \alpha$ for all i. These cases will be examined after the next paragraph.

An arbitrary interval assignment $F(x) = [x^-, x^+]$ is a valid representation for the interval order shown in Fig. 9.9 if and only if $x_0^- < x_0^+, x_{k+1}^- < x_{k+1}^+$,

Discontinuities and Admissible Lengths

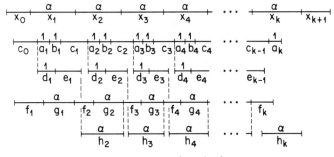

Figure 9.9 $\theta(X, \prec) = \left(2 - \frac{1}{k}, 2\right) \cup (k, \infty)$.

A. $\{x_0^-, x_1^-, c_0^-, f_1^-\} \le x_0^+ < \{a_1^-, d_1^-\} \le c_0^+ < g_1^- \le f_1^+ < b_1^- \le a_1^+ <$
   $e_1^- \le d_1^+ < c_1^- \le b_1^+ < \{x_2^-, f_2^-\}$.

B. $\{x_i^-, f_i^-\} \le \{h_{i-1}^+, g_{i-1}^+, x_{i-1}^+\} < h_i^- \le e_{i-1}^+ < \{a_i^-, d_i^-\} \le c_{i-1}^+ < g_i^-$
   $< f_i^+ < b_i^- \le a_i^+ < e_i^- \le d_i^+ < c_i^- \le b_i^+ < \{x_{i+1}^-, f_{i+1}^-\}$ *for* $2 \le i \le k$
   $- 1$ (*there is no* h_1^+).

C. $\{x_k^-, f_k^-\} \le \{h_{k-1}^+, g_{k-1}^+, x_{k-1}^+\} < h_k^- \le e_{k-1}^+ < a_k^- \le \{c_{k-1}^+, f_k^+, a_k^+\}$
   $< x_{k+1}^- \le \{x_k^+, h_k^+, x_{k+1}^+\}$.

We shall use these to validate specific two-lengths assignments.

CASE 1. $\rho(e_i) = 1$ for all i. Then $\alpha < 2$. Moreover, since $\Sigma_1^k \rho(x_i) > \Sigma_1^k \rho(a_i)$ $+ \Sigma_1^{k-1} \rho(e_i)$, $\alpha k > 2k - 1$, or $\alpha > 2 - 1/k$. Hence $2 - 1/k < \alpha < 2$ for Case 1. We assign specific intervals to all points, with all c_i, e_i, and f_i short, to show that all such α are admissible.

With λ positive and small, the assignment is

$F(x_i) = [\alpha(i - 1), \alpha i]$ $0 \le i \le k + 1$

$F(a_i) = [2i - 2 + \lambda(2i - 1), 2i - 1 + \lambda(2i - 1)]$ $1 \le i \le k$

$F(b_i) = [2i - \alpha + \lambda(2i), 2i + 1 - \alpha + \lambda(2i)]$ $1 \le i \le k - 1$

$F(c_0) = [-1 + 2\lambda, 2\lambda]$

$F(c_i) = [2i - 1 + \lambda 2(i + 1), 2i + \lambda 2(i + 1)]$ $1 \le i \le k - 1$

$F(d_i) = [2i - 2 + \lambda(2i), 2i - 1 + \lambda(2i)]$ $1 \le i \le k - 1$

$F(e_i) = [2i - 1 + \lambda(2i), 2i + \lambda(2i)]$ $1 \le i \le k - 1$

$F(f_i) = [2i - 1 - \alpha + \lambda(2i - 1), 2i - \alpha + \lambda(2i - 1)]$ $1 \le i \le k$

$F(g_i) = [2i - \alpha + \lambda(2i - 1), 2i + \lambda(2i - 1)]$ $1 \le i \le k - 1$

$F(h_i) = [2i - 2 + \lambda(2i - 2), 2i - 2 + \alpha + \lambda(2i - 2)]$ $2 \le i \le k$.

The x_i, g_i, and h_i have length α. All others have length 1.

It is straightforward but tedious to show that A, B, and C hold for this assignment. A summary follows. The critical inequality for A is $c_1^- \le b_1^+$, or $1 + 4\lambda \le 3 - \alpha + 2\lambda$, which implies $\lambda \le (2 - \alpha)/2$. Given this, B requires $\alpha \ge (2 - 1/i)(1 + \lambda)$ for $f_i^- \le x_{i-1}^+$, and $\alpha > (2i + 1 + \lambda 2i)/(i + 1)$ for $b_i^+ < x_{i+1}^-$. The tightest cases occur here when $i = k - 1$; they require

$$\lambda \le \left[\alpha - \left(2 - \frac{1}{k}\right) + \frac{1}{k(k-1)}\right]\left(\frac{k-1}{2k-3}\right),$$

$$\lambda < \left[\alpha - \left(2 - \frac{1}{k}\right)\right]\left(\frac{k}{2(k-1)}\right),$$

which hold for small λ. The critical inequality for C is $f_k^- \le x_{k-1}^+$, or

$$\lambda \le \left[\alpha - \left(2 - \frac{1}{k}\right)\right]\left(\frac{k}{2k-1}\right).$$

Since $\alpha > (2 - 1/k)$, all of A through C hold for small λ.

CASE 2. $\rho(e_i) = \alpha$ for all i. Then, since $\Sigma_1^k \rho(x_i) > \Sigma_1^k \rho(a_i) + \Sigma_1^{k-1} \rho(e_i)$, $\alpha k > \alpha(k - 1) + k$, or $\alpha > k$. Hence $k < \alpha < \infty$ for Case 2. Our interval assignment for this case is

$F(x_i) = [\alpha(i - 1), \alpha i]$ $\qquad 0 \le i \le k + 1$

$F(a_i) = [(\alpha + 1)(i - 1) + \lambda(2i - 1), (\alpha + 1)(i - 1) + 1 + \lambda(2i - 1)]$
$\qquad\qquad\qquad\qquad\qquad\qquad\qquad\qquad 1 \le i \le k$

$F(b_i) = [(\alpha + 1)(i - 1) + 1 + \lambda(2i - 1), (\alpha + 1)(i - 1) + 2 + \lambda(2i - 1)]$
$\qquad\qquad\qquad\qquad\qquad\qquad\qquad\qquad 1 \le i \le k - 1$

$F(c_i) = [\alpha(i - 1) + i + 1 - \lambda, \alpha i + i + 1 - \lambda]$ $\qquad 0 \le i \le k - 1$

$F(d_i) = [(\alpha + 1)(i - 1) + \lambda(2i), (\alpha + 1)(i - 1) + 1 + \lambda(2i)]$
$\qquad\qquad\qquad\qquad\qquad\qquad\qquad\qquad 1 \le i \le k - 1$

$F(e_i) = [(\alpha + 1)(i - 1) + 1 + \lambda(2i), (\alpha + 1)i + \lambda(2i)]$
$\qquad\qquad\qquad\qquad\qquad\qquad\qquad\qquad 1 \le i \le k - 1$

$F(f_i) = [(\alpha + 1)(i - 2) + 2 + \lambda(2i - 2),$
$\qquad\qquad\qquad (\alpha + 1)(i - 2) + 2 + \alpha + \lambda(2i - 2)] \quad 1 \le i \le k$

$F(g_i) = [\alpha(i - 1) + i, \alpha(i - 1) + i + \alpha]$ $\qquad 1 \le i \le k - 1$

$F(h_i) = [(\alpha + 1)(i - 1) + \lambda(2i - 2), (\alpha + 1)(i - 1) + \alpha + \lambda(2i - 2)]$
$\qquad\qquad\qquad\qquad\qquad\qquad\qquad\qquad 2 \le i \le k.$

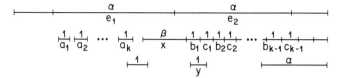

Figure 9.10 $\theta(X, \prec) = (k, 2k - 1) \cup (2k - 1, \infty)$.

A detailed analysis shows that this assignment satisfies A, B, and C when $\alpha > k$ and $0 < \lambda < \min\{1, \alpha - k\}/2k$. Hence all $\alpha > k$ are admissible for Case 2. □

Proof of Theorem 8. We consider the interval order pictured in Fig. 9.10 with free interval x of length β.

CASE 1. $\beta = 1$. Then

$$\alpha = \rho(e_2) < \rho(x) + \sum_{i=1}^{k-1}[\rho(b_i) + \rho(c_i)] = 2k - 1,$$

so $\alpha < 2k - 1$. Moreover, $\alpha = \rho(e_1) > \Sigma\rho(a_i) = k$, and $2\alpha = \rho(e_1) + \rho(e_2) > \Sigma\rho(a_i) + \rho(x) + \rho(y) + \Sigma_2^{k-1}\rho(b_i) = 2m$, so that $\alpha > k$. Hence $k < 2 < 2k - 1$ for Case 1.

All values of α in $(k, 2k - 1)$ are admissible. If α is near k, the a_i must be tightly packed and nonintersecting, along with $x, y, b_2, \ldots, b_{k-1}$, and $b_1, c_1, \ldots, c_{k-1}$ must overlap their neighbors substantially. Moreover, the point where e_1 and e_2 touch is positioned near the left endpoint of x.

On the other hand, if α is near $2k - 1$, the right part of the second line is less densely packed, much as shown in Fig. 9.10. For α very near to $2k - 1$, the left endpoint of e_2 is near the left endpoint of x, and the right endpoint of e_2 is near the right endpoint of c_{k-1}.

CASE 2. $\beta = \alpha$. Then $\alpha > 2k - 1$ since $\rho(e_1) + \rho(e_2) > \rho(x) + \Sigma\rho(a_i) + \rho(y) + \Sigma_2^{k-1}\rho(b_i)$. For α near $2k - 1$, the a_i are tightly packed along with x, y, and the b_i for $i \geq 2$, much as for α near k in Case 1. Since $\rho(x) = \alpha$ here, the point where e_1 and e_2 touch is placed slightly to the left of the center of x since there is one more a_i to pack between the left endpoint of e_1 and the left endpoint of x than there are $y, b_2, \ldots, b_{k-1}$ to pack between the right endpoint of x and the right endpoint of e_2.

There is no upper bound on α for Case 2. When α is large, the right endpoint of e_2 is positioned in a suitable way, and the rest of the construction is straightforward. □

10

Extremization Problems

Our final chapter, like Chapter 6, is devoted to extremization problems. It addresses two questions for each positive integer k. First, what is the cardinality of the smallest interval order that cannot be represented by real intervals that have fewer than k different lengths? Second, for each $m \geq 2k - 1$, what is the cardinality of the smallest magnitude-m interval order that cannot be represented with fewer than k lengths? Exact answers are known only in special cases. Some bounds on the answers are obtained for the general cases, and a number of conjectures are proposed.

10.1 DEFINITIONS AND CONJECTURES

Throughout this chapter we maintain the assumptions that X is a nonempty finite set and that all intervals used in representations have positive lengths. In addition, with no loss in generality, it will be assumed that no two distinct points x and y in an interval order $(X, \prec)$ are equivalent in the sense of $x \approx y$ (Sections 1.3, 2.1). It follows that every interval order $(X, \prec)$ is uniquely described by its characteristic matrix $\mathbf{M}(X, \prec)$. When $(X, \prec)$ has magnitude m, the domain of $\mathbf{M}$ is $\{(i, j): 1 \leq i \leq j \leq m\}$ and its values are

$\mathbf{M}_{ij} = 1$ if some $x \in X$ is in the ith class in $(X/\sim^-, \prec^-)$ and the jth class in $(X/\sim^+, \prec^+)$;

$\mathbf{M}_{ij} = 0$ otherwise.

As noted in Section 2.3, each such matrix has a 1 in every row and in every column, and every upper-triangular $m \times m$ 0-1 matrix with this property is the characteristic matrix of a magnitude-m interval order.

We shall focus on two functions. The first function, σ, is defined on $\{1, 2, \ldots\}$ for interval orders $(X, \prec)$ by

$$\sigma(k) = \min_{(X, \prec)} \{|X|: \rho^*(X, \prec) \geq k\}.$$

Definitions and Conjectures

Thus $\sigma(k)$ is the smallest $|X|$ that has an interval order all of whose real representations require k or more different interval lengths. Clearly $\sigma(1) = 1$. For $k \geq 2$, since $|X| < \sigma(k)$ implies $\rho^*(X, \prec) < k$,

$$\sigma(k) = \max\{n: |X| < n \Rightarrow \rho^*(X, \prec) \leq k - 1\}.$$

Alternatively, $\sigma(k)$ is the smallest $|X|$ such that some interval order on X is in $\mathcal{P}_k$ but not $\mathcal{P}_{k-1}$; and the maximum n such that every n-point interval order can be represented with k or fewer lengths is $\sigma(k+1) - 1$.

The second focal function, ν, is defined on

$$K^* = \{(k, m): k \in \{1, 2, \dots\} \text{ and } m \geq 2k - 1\}$$

by

$$\nu(k, m) = \min_{(X, \prec)} \{|X|: (X, \prec) \text{ has magnitude } m, \text{ and } \rho^*(X, \prec) \geq k\},$$

so that $\nu(k, m)$ is the smallest magnitude-m interval order all of whose representations require at least k interval lengths. The reason for K^* is that if $m < 2k - 1$, then every magnitude-m interval order has a representation that uses fewer than k lengths, but some magnitude-$(2k - 1)$ interval orders, including ψ_m of Section 5.5, require k lengths. See, for example, Theorem 1 in the next section.

Our two functions are related by

$$\sigma(k) = \min_m \{\nu(k, m): m \geq 2k - 1\}$$

since every $(X, \prec)$ that requires k or more lengths must have $m(X, \prec) \geq 2k - 1$.

The results established in this chapter are motivated by two conjectures that apply to all $k \in \{1, 2, \dots\}$:

Conjecture 1. $\sigma(k) = 3k - 2$.

Conjecture 2. $\nu(k, m)$ is nondecreasing in $m \geq 2k - 1$.

Theorem 5.18, which says that a depth-k interval order on $3k - 2$ points has $\rho^*(X, \prec) = k$, implies that $\sigma(k) \leq 3k - 2$. It is not hard to show that $\sigma(2) = 4$ and $\sigma(3) = 7$. Moreover, these values and a simple induction on k which removes extreme end intervals shows that $\rho^*(X, \prec) \leq k$ whenever $|X| = 2(k + 1)$ and $k \geq 2$, so that $\sigma(k) \geq 2k$ for all $k \geq 2$. Thus we know that

$$2k \leq \sigma(k) \leq 3k - 2 \quad \text{for} \quad k \geq 2,$$

but little more is known about σ at the present time. Despite this, the efficiency of the construction in Fig. 5.6 for forcing different lengths makes it seem virtually certain that Conjecture 1 is true.

Conjecture 2 is stronger than Conjecture 1 since it implies Conjecture 1 by way of $\sigma(k) = \min\{\nu(k, m)\}$ and $\nu(k, 2k - 1) = 3k - 2$, the latter of which is proved in Section 10.3. As we shall see in Section 10.4, $\nu(k, m)$ is not strictly increasing in $m \geq 2k - 1$ since there are $(k, m) \in K^*$ at which $\nu(k, m) = \nu(k, m + 1)$. The question of whether $\nu(k, m)$ can decrease at some m remains open.

Apart from the last section of the chapter, which discusses additional conjectures for σ and ν, the ensuing sections concentrate on ν. The next section presents three basic results for ν, including

$$\nu(k, m) \leq m + k - 1.$$

Section 10.3 then shows that this is actually an equality for $k \leq 4$ and for m near $2k - 1$. Equality fails at $k = 5$ as we shall demonstrate in Section 10.4. That section also considers the limiting behavior in m of $\nu(k, m) - m$ and shows that $(m + k - 1) - \nu(k, m)$ is unbounded over K^*.

Some of our results apply also to $\tilde{\sigma}$ and $\tilde{\nu}$, defined analogously for interval graphs $(X, \sim)$ by

$$\tilde{\sigma}(k) = \min_{(X, \sim)} \{|X|: \rho^*(X, \sim) \geq k\},$$

$$\tilde{\nu}(k, m) = \min_{(X, \sim)} \{|X|: (X, \sim) \text{ has magnitude } m, \text{ and } \rho^*(X, \sim) \geq k\},$$

where $\rho^*(X, \sim)$ is defined as in Section 9.1. For example, since

$$\sigma(k) \leq \min_{(X, \sim)} \{|X|: \rho^*(X, \prec) \geq k \text{ for every } (X, \prec) \text{ that agrees with } (X, \sim)\}$$
$$= \min_{(X, \sim)} \{|X|: \rho^*(X, \sim) \geq k\} = \tilde{\sigma}(k),$$

we have $2k \leq \sigma(k) \leq \tilde{\sigma}(k)$ for all $k \geq 2$. Moreover, since there are no buried subgraphs in the construction used for Theorem 5.18 when end intervals are extended, it follows from Theorem 3.11 that $\tilde{\sigma}(k) \leq 3k - 2$. Therefore

$$2k \leq \tilde{\sigma}(k) \leq 3k - 2 \quad \text{for} \quad k \geq 2.$$

In addition, the proofs of Theorems 1 and 2 in the next section adapt readily to $\tilde{\nu}$ via Theorem 3.11, so that

$$\tilde{\nu}(k, m) \leq m + k - 1 \quad \text{for} \quad (k, m) \in K^*.$$

It is not presently known whether σ and $\tilde{\sigma}$, or ν and $\tilde{\nu}$, differ for any argument.

10.2 BASIC THEOREMS

In this section we prove three basic theorems on the definition and behavior of ν. The first theorem concerns the domain of ν.

Theorem 1. *Suppose m is a positive integer. Then there is a magnitude-m interval order $(X, \prec)$ for which $\rho^*(X, \prec) \geq k$ if and only if $k \leq \lfloor (m+1)/2 \rfloor$.*

It follows from Theorem 1 that the domain of definition of ν is

$$K^* = \left\{ (k,m) : m \in \{1, 2, \ldots\} \text{ and } k \in \left\{1, \ldots, \left\lfloor \frac{m+1}{2} \right\rfloor \right\} \right\}$$

$$= \{(k,m) : k \in \{1, 2, \ldots\} \text{ and } m \in \{2k-1, 2k, \ldots\}\}.$$

Theorem 2. $\nu(k, m) \leq m + k - 1$ *for all $(k, m) \in K^*$.*

Theorem 3. $\nu(k, m) + 2 \leq \nu(k+1, m+2) \leq \nu(k, m) + 3$ *for all $(k, m) \in K^*$.*

Equality in Theorem 2 is discussed in the next section, and strict inequality cases are presented in Section 10.4. The latter section's results imply that the first inequality in Theorem 3 is sometimes an equality: there are (k, m) for which $\nu(k, m) + 2 = \nu(k+1, m+2)$. Theorem 3 can also be used (but is not essential) in conjunction with the theorem in the next section to conclude that Conjecture 1 is true whenever $k \leq 7$.

Proofs of Theorems 1 and 2. Given m, let $\mathbf{M}_1$ be the $m \times m$ characteristic matrix with 1's on the main diagonal and 0's elsewhere. Its interval order is an m-point chain. Add 1's to $\mathbf{M}_1$, beginning in the upper right corner at cell $(1, m)$, and proceed southwest diagonally until $\lfloor (m-1)/2 \rfloor$ 1's have been added. These additions successively require $2, 3, \ldots, \lfloor (m-1)/2 \rfloor$ lengths (by $\subset_0$ chains) to represent the corresponding interval orders. Thus we can force $\lfloor (m+1)/2 \rfloor$ lengths at magnitude m, and Theorem 2 follows immediately from the construction since k lengths are forced by $m + k - 1$ points.

The proof of Theorem 1 is completed by showing that every magnitude-m interval order is representable with $\lfloor (m+1)/2 \rfloor$ or fewer lengths. It suffices to consider the most demanding case, that is, ψ_m, whose characteristic matrix has a 1 in every cell.

Partition the cells of $\mathbf{M}$ in the manner shown in Fig. 10.1. We shall assign the same interval length to every cell in each part of the partition in such a way that the resulting interval assignment represents ψ_m. This will be done for odd m. Since there are $(m+1)/2$ parts in the partition for odd m, it follows that $\rho^*(\psi_m) = (m+1)/2$. The desired result for even m then obtains by removing the main diagonal for $m + 1$ since this decreases by 1 the number of lengths used.

Given odd $m \geq 3$, let $\mu = (m + 1)/2$, and let

$$c_i = 4^i \quad \text{for} \quad i = 0, 1, \ldots, \mu - 1.$$

Also let $F(i, j)$ denote the interval assigned to the point in ψ_m that corresponds to cell (i, j) in **M**. The μ intervals along the opposite diagonal ↗ are

$$F(\mu - i, \mu + i) = [-c_i, c_i] \quad 0 \leq i \leq \mu - 1.$$

The other intervals are

$$F(i, j) = [-c_{\mu-j} - 2c_{\mu-i}, -c_{\mu-j}] \qquad i < \mu, j \leq \mu;$$
$$F(i, j) = [c_{j-\mu} - 2c_{\mu-i}, c_{j-\mu}] \qquad i < \mu < j < 2\mu - i;$$
$$F(i, j) = [-c_{\mu-i}, -c_{\mu-i} + 2c_{j-\mu}] \qquad i < \mu, j > 2\mu - i;$$
$$F(i, j) = [c_{i-\mu}, c_{i-\mu} + 2c_{j-\mu}] \qquad i \geq \mu, j > \mu.$$

The length of each row-i interval up to the opposite diagonal (where $i + j = 2\mu$) is $2c_{\mu-i}$, and the length of each column-j interval below the opposite diagonal is $2c_{j-\mu}$. These lengths satisfy the pattern indicated in Fig. 10.1.

To show that F is a valid representation of ψ_m, consider $F(i, j)$ and $F(p, q)$ with $i \leq p$ for definiteness. We have $(i, j) \prec (p, q)$ if $j < p$, and $(i, j) \sim (p, q)$ otherwise with regard to ψ_m. We consider these in turn.

CASE 1. $j < p$. In this case, $F(i, j)$ must end before $F(p, q)$ begins. Fix p. Given $j < p$, the right endpoint of $F(i, j)$ for fixed i is maximized at $j = p - 1$; then, with j fixed at $p - 1$, the right endpoint of $F(i, p - 1)$ is maximized at $i = p - 1$. Similarly, with p still fixed, the left endpoint of $F(p, q)$ is minimized at $q = p$. Hence the worst-case situation is $F(p - 1,$

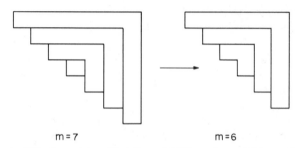

Figure 10.1 Same interval length used within each part of the partition.

Basic Theorems

$p - 1$) versus $F(p, p)$, for intervals along the main diagonal. By the definitions,

$$F(p, p) = [-3c_{\mu-p}, -c_{\mu-p}] \quad \text{for} \quad p < \mu$$

$$F(\mu, \mu) = [-c_0, c_0]$$

$$F(p, p) = [c_{p-\mu}, 3c_{p-\mu}] \quad \text{for} \quad p > \mu.$$

Since $c_i = 4^i$, $\sup F(p, p) < \inf F(p + 1, p + 1)$ for $p \leq \mu - 1$.

CASE 2. $p \leq j$. In this case, $F(i, j)$ and $F(p, q)$ must intersect for a valid representation. We show first that $\inf F(i, j) \leq \sup F(p, q)$. With p fixed, the left endpoint of $F(i, j)$ is maximized, when j is fixed, at $i = p$. The worst case here comes down to $F(p, p)$ versus $F(p, m)$:

$$F(1, m) = [-c_{\mu-1}, c_{\mu-1}]$$

$$F(p, m) = [-c_{\mu-p}, -c_{\mu-p} + 2c_{m-\mu}] \quad \text{for} \quad 1 < p < \mu$$

$$F(p, m) = [c_{p-\mu}, c_{p-\mu} + 2c_{m-\mu}] \quad \text{for} \quad p \geq \mu.$$

When $p < \mu$, $F(p, m) \cap F(p, p) = \{-c_{\mu-p}\}$; $F(\mu, m) \cap F(\mu, \mu) = \{c_0\}$; for $p > \mu$, $F(p, m)$ and $F(p, p)$ begin at the same point.

We show next that $\inf F(p, q) \leq \sup F(i, j)$. With p fixed, the right endpoint of $F(i, j)$ for fixed i is minimized at $j = p$, and the left endpoint of $F(p, q)$ is maximized at $q = p$. The worst case is $F(1, p)$ versus $F(p, p)$. This is symmetric to the case at the end of the preceding paragraph, so $F(1, p) \cap F(p, p) \neq \emptyset$ for all p. □

Proof of Theorem 3. We show first that $\nu(k + 1, m + 2) \leq \nu(k, m) + 3$. Let $\mathbf{M}_0$ be an $m \times m$ characteristic matrix. Expand $\mathbf{M}_0$ by adding a new top row and a new right column to make an $(m + 2) \times (m + 2)$ matrix with 1's in the three corners of the expanded matrix: see the top matrix in Fig. 10.2. The expansion clearly forces a new longest length, so if $\nu(k, m)$ 1's in $\mathbf{M}_0$ is the fewest that will force k lengths there, then $\nu(k, m) + 3$ or fewer 1's will force $k + 1$ lengths in the expansion. Hence $\nu(k + 1, m + 2) \leq \nu(k, m) + 3$.

To prove that $\nu(k, m) + 2 \leq \nu(k + 1, m + 2)$, suppose first that $\mathbf{M}'$ is a magnitude-m' characteristic matrix with the fewest 1's, say $\nu(k', m')$, that force k' lengths in an interval representation. Suppose further that $\mathbf{M}'$ has three or more 1's in its first row: see the middle matrix of Fig. 10.2. A nonextreme 1 in the first row can always be assigned the same length as the right-most 1 in that row by left extension if necessary. The only crucial aspect of the nonextreme 1's interval is its right endpoint, which can be preserved intact if this 1 is moved down its column (see arrow) to occupy a cell in a lower row. Such a move cannot decrease the number of lengths needed and would increase the

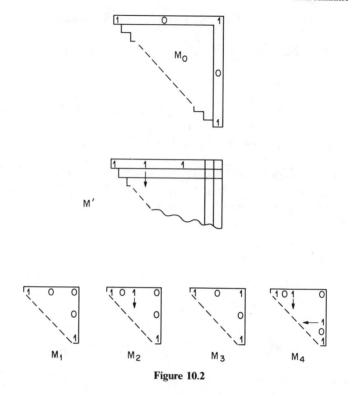

Figure 10.2

number of lengths by one were it not for the assumption that **M'** is optimal at (k', m').

It follows that some **M'** that is optimal at (k', m') has at most two 1's in its first row and, by analogy, at most two 1's in its last column.

According to this, we assume that an optimal **M** for $(k + 1, m + 2)$, with $v(k + 1, m + 2)$ 1's that force $k + 1$ lengths, has one of the four forms at the bottom of Fig. 10.2. For each of these, we shall delete the first row and last column, moving noncorner 1's inward (see arrows) so that the reduced $m \times m$ matrix is a characteristic matrix with a 1 in every row and column.

M$_1$. Since the 1's at the ends of the main diagonal can be assigned the same length as a 1 elsewhere, the reduction gives $v(k, m) \leq v(k + 1, m) \leq v(k + 1, m + 2) - 2$.

M$_2$. When the second 1 in the first row is moved down, it could decrease the number of lengths needed by one, but not more (else its move back up to the first row would imply that fewer than $k + 1$ lengths suffice for **M**$_2$). Since two 1's vanish in the reduction, $v(k, m) \leq v(k + 1, m + 2) - 2$.

M$_3$. Since the 1 in the upper right corner forces a longest length, optimality of **M**$_3$ for $(k + 1, m + 2)$ implies that $v(k, m) = v(k + 1, m + 2) - 3$.

M$_4$. When the second 1 in the first row and the first 1 in the last column are moved inward, the number of lengths needed cannot decrease by more than

Exact Values

one (else $\mathbf{M}_4$ would not be optimal). Again, as in $\mathbf{M}_2$, $\nu(k, m) \leq \nu(k + 1, m + 2) - 2$.

Hence $\nu(k, m) + 2 \leq \nu(k + 1, m + 2)$ in all cases. □

Remark 1. If $\mathbf{M}_4$ is an optimal array but $\mathbf{M}_3$ is not, then $\mathbf{M}_4$ must have 0's beneath the first-row 1 that is moved down, and 0's to the left of the last-column 1 that is moved to the left. Moreover, the two 1's that are moved cannot be placed in the same cell, else the addition of 1 in the upper right corner would given an optimal $\mathbf{M}_3$.

10.3 EXACT VALUES

Our next theorem identifies a number of situations in which the inequality of Theorem 2 is an equality.

Theorem 4. *Suppose $(k, m) \in K^*$. Then $\nu(k, m) = m + k - 1$ if either of the following holds*:
  (a) $k \leq 4$,
  (b) $m \in \{2k - 1, 2k, 2k + 1\}$.

The proof is rather long and will take up the rest of this section. We begin with the proof of (b) since it interacts significantly with the preceding proofs. Each of the (b) and (a) proofs is divided into two parts.

Proof of (b) *for* $m \in \{2k - 1, 2k\}$. Both values of m are accounted for by the claim that $\nu(\lfloor(m + 1)/2\rfloor, m) = \lfloor(3m - 1)/2\rfloor$: if m is odd, k here is $(m + 1)/2$ with $m = 2k - 1$; if m is even, k is $m/2$ with $m = 2k$. Before proving this claim, we show that there is a minimum-lengths representation of ψ_m that assigns the same interval length to every cell in each part of the partition of $\mathbf{M}$ shown in Fig. 10.3. This provides a different proof of Theorem 1 than the one using Fig. 10.1. We verify this for odd m only since the result for even m then follows by deleting the first row of the matrix for ψ_{m+1}.

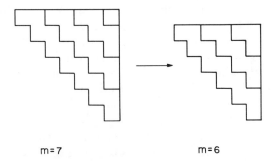

m = 7   m = 6

Figure 10.3 Same interval length used within each part of the partition.

Given odd $m \geq 3$, let N be large compared to m. We define a representation F for ψ_m by

$$F(i, j) = [Ni, Nj] \qquad \text{if } j - i \text{ is odd,}$$
$$F(i, j) = [(N + 1)(i - 1), Nj + i - 1] \qquad \text{if } j - i \text{ is even.}$$

The lengths are $N(j - i)$ for odd $j - i$, and $N(j - i + 1)$ for even $j - i$. Since each part of the partition in Fig. 10.3 consists of all (i, j) with $j - i \in \{c, c + 1\}$ for even c, all intervals in a given part have the same length.

As in the proof of Theorem 1, we compare $F(i, j)$ and $F(p, q)$ with $i \leq p$ and p fixed. Suppose first that $j < p$, which requires $(i, j) \prec (p, q)$. The left endpoint of $F(p, q)$ is Np or $(N + 1)(p - 1)$, and both of these exceed $(N + 1)(p - 1) - 1$, which is the maximum right endpoint of the $F(i, j)$ with $i \leq j < p$. Suppose next that $i \leq p \leq j$, which requires $(i, j) \sim (p, q)$. Then the right endpoint of $F(i, j)$ is minimized at Np, and the left endpoint of $F(p, q)$ is at Np or $(N + 1)(p - 1)$. Also, the right endpoint of $F(p, q)$ is minimized at Np, and the left endpoint of $F(i, j)$ is maximized at Np or $(N + 1)(p - 1)$. Since $Np > (N + 1)(p - 1)$ for large N, $F(i, j) \cap F(p, q) \neq \emptyset$.

We now prove the claim that $\nu((m + 1)/2, m) = (3m - 1)/2$ for odd m. Let $\mathbf{M}$ be an optimal $m \times m$ characteristic matrix that forces $(m + 1)/2$ lengths. Then $\mathbf{M}$ has 1 in the upper right corner, else the result just proved shows that $(m - 1)/2$ or fewer lengths suffice. By the argument in the second and third paragraphs of the proof of Theorem 3, $\mathbf{M}$ can be assumed to have form $\mathbf{M}_3$ in Fig. 10.2. Given this form, delete the first row and last column. The reduced $(m - 2) \times (m - 2)$ matrix $\mathbf{M}'$ must force $(m - 1)/2$ lengths. By the argument used earlier in this paragraph for $\mathbf{M}$, $\mathbf{M}'$ has 1 in its upper right corner. Moreover, since $\mathbf{M}'$ must be optimal for $((m - 1)/2, m - 2)$ when $\mathbf{M}$ is optimal for $((m + 1)/2, m)$, $\mathbf{M}'$ can be assumed to have form $\mathbf{M}_3$. By continuation, we conclude that $\mathbf{M}$ has 1's in the $(m + 1)/2$ cells along its opposite diagonal. In addition, it has $m - 1$ more 1's for its first $(m - 1)/2$ columns and its last $(m - 1)/2$ rows. The total number of 1's in $\mathbf{M}$ is therefore $(m + 1)/2 + m - 1 = (3m - 1)/2$.

To prove the claim that $\nu(m/2, m) = (3m - 2)/2$ for even m, assume that $\mathbf{M}$ is an optimal $m \times m$ characteristic matrix that forces $m/2$ lengths. By the initial partition result in this proof, $\mathbf{M}$ has 1 in one of cells $(1, m - 1)$, $(1, m)$, and $(2, m)$, which constitute the upper right part of the partition. Assume for the time being that $m \geq 4$. By the proof of Theorem 3, $\mathbf{M}$ can be assumed to have form $\mathbf{M}_2$, $\mathbf{M}_3$, or $\mathbf{M}_4$ in Fig. 10.2. Suppose it cannot have form $\mathbf{M}_3$, so it has 0 in cell $(1, m)$, and assume for definiteness that $\mathbf{M}_{1, m-1} = 1$. Then Remark 1 at the end of the preceding section implies that $\mathbf{M}$ doesn't have form $\mathbf{M}_4$, hence that it has form $\mathbf{M}_2$ with $\mathbf{M}_{m-1, m-1} = 1$ since the only 1 in the last column is in the lower right corner. However, because $\mathbf{M}_{m-1, m-1} = 1$, the 1 in cell $(1, m - 1)$ can be moved to cell $(1, m)$ without decreasing the number of lengths needed or changing the number of 1's. But then $\mathbf{M}_3$ is an optimal $\mathbf{M}$.

Exact Values

Therefore, when $m \geq 4$, an optimal **M** for $(m/2, m)$ can be assumed to have form $\mathbf{M}_3$. When the first row and last column are deleted, a similar result holds for the reduced matrix so long as $m - 2 \geq 4$. Continuation shows that some optimal **M** has 1's in each of the $m/2 - 1$ cells on the opposite diagonal from $(1, m)$ to $(m/2 - 1, m/2 + 2)$ inclusive. These force $m/2 - 1$ lengths. The fewest additional 1's needed to cover the first $m/2 + 1$ columns and the last $m/2 + 1$ rows is m. All of these can be put along the main diagonal, where the middle two stand in the relation C_0 to the 1 in cell $(m/2 - 1, m/2 + 2)$, and this forces the final length. Hence an optimal **M** for $(m/2, m)$ has exactly $(m/2 - 1) + m = (3m - 2)/2$ cells with 1's. □

Proof of (b) *for* $m = 2k + 1$. We are to show that for odd $m \geq 3$, $\nu((m - 1)/2, m) = (3m - 3)/2$. Since $\nu(1, 3) = 3$, the result holds for $m = 3$. Given $m \geq 5$ and odd, we assume the result holds for odd $m' < m$ and prove that it holds also for m.

Let **M** be an optimal $m \times m$ characteristic matrix for $(k, m) = ((m - 1)/2, m)$. Suppose first that **M** can have form $\mathbf{M}_3$ in Fig. 10.2. Then, when its first row and last column are deleted, what remains must be optimal for $((m - 3)/2, m - 2)$ with $\nu((m - 3)/2, m - 2) + 3 = \nu((m - 1)/2, m)$. By the induction hypothesis, we then have $\nu((m - 1)/2, m) = (3m - 9)/2 + 3 = (3m - 3)/2$.

Suppose henceforth that no $\mathbf{M}_3$ form is optimal for $((m - 1)/2, m)$. Then a form like $\mathbf{M}_1$, $\mathbf{M}_2$, or $\mathbf{M}_4$ is optimal. If $\mathbf{M}_1$ or $\mathbf{M}_2$ applies, remove the last column, where the only 1 is at the bottom. The reduced matrix is optimal for $(m - 1)/2$ lengths at magnitude $m - 1$, and therefore, by the preceding (b) proof, $\nu((m - 1)/2, m - 1) = (3m - 5)/2$. Hence **M** has $(3m - 5)/2 + 1$ cells with 1's, so $\nu((m - 1)/2, m) = (3m - 3)/2$.

Finally, suppose only $\mathbf{M}_4$ can be an optimal form for **M** among $\mathbf{M}_1$ through $\mathbf{M}_4$. We show that this is impossible, which completes the present proof. Since **M** forces $(m - 1)/2$ lengths and has $\mathbf{M}_{1,m} = 0$, the equal-lengths assignment used with Fig. 10.3 implies that $\mathbf{M}_{ij} = 1$ for some

$$(i, j) \in \{(1, m - 2), (1, m - 1), (2, m - 1), (2, m), (3, m)\},$$

since otherwise $(m - 3)/2$ lengths would suffice. This fact and Remark 1 at the end of the preceding section imply that **M** (with form $\mathbf{M}_4$) looks like one of the two matrices in Fig. 10.4. Partial interval representations appear beneath the matrices.

Suppose the left matrix obtains. Then one less length is needed if 1_a, in cell $(m - 1, m)$, is moved to cell $(m - 1, m - 1)$, labeled as x on the figure, since otherwise form $\mathbf{M}_2$ would be optimal. However, this move does not in fact decrease the number of lengths required. Initially, 1_a could have the longest length used for other intervals, and x could have the same length and same endpoints as 1_a after the move, with the interval for cell (m, m) moved right to avoid intersection with x.

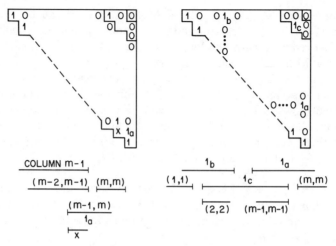

Figure 10.4 Elaborations of M_4.

Suppose then that the right matrix obtains. We can assume that 1_a, 1_b, and 1_c use the same longest length in a minimum-lengths representation of **M**. The assumption that only form M_4 is optimal then implies that the number of lengths needed must decrease if 1_a is moved left, or if 1_b is moved down. But in fact such moves will not decrease the number of lengths needed. For example, since 1_a has the same length as 1_c, only its left endpoint is crucial in the placements. This endpoint can remain fixed when 1_a is moved left in its row, but the right endpoint of the interval for 1_a thus moved may put new restrictions on other placements, and these can only increase the number of lengths needed if this number changes. □

Proof of (a) *for* $k = 3$. Since the cases for $k \leq 2$ are obvious, we consider $k = 3$. It will be shown that there is no way to force more than two lengths with an $m \times m$ characteristic matrix **M** that has $(m + 1)$ 1's, so, by Theorem 2, $\nu(3, m) = m + 2$ when $m \geq 5$.

Given $(m + 1)$ 1's in **M**, the only way one might force more than two lengths is to have some 0's on the main diagonal. Suppose there are $(m - h)$ 1's and $h > 0$ 0's on the main diagonal. Each 0 must have a 1 above it and another 1 to its right. Therefore, to have only $h + 1$ off-diagonal 1's (for a total of $m + 1$), successive 0's on the main diagonal must share a 1 in common, which is in the row of the first 0 and the column of the second.

This is pictured in Fig. 10.5. The off-diagonal 1's are in cells $(i_1, i_2), (i_2, i_3), \ldots, (i_h, i_{h+1}), (i_{h+1}, i_{h+2})$ with $i_1 < i_2 < \cdots < i_{h+2}$, and with $\sim$ between successive cells and $\prec$ between nonsuccessive cells. I shall refer to such a sequence as a *strict path*. In terms of Section 1.7, it is a $\sim$-path with no chord.

The strict path configuration with $(m + 1)$ 1's in **M** does not force more than two lengths. The off-diagonal 1's in the strict path can be assigned the same length with successive intervals touching end-to-end. A shorter length is

Exact Values

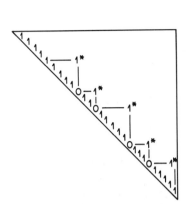

Figure 10.5 A strict path ($*$).

then used for the 1's on the main diagonal to position their chain among the longer intervals. □

Proof of (a) *for k* = 4. Let **M** be a magnitude-m characteristic matrix with $m \geq 7$. To prove that $v(4, m) = m + 3$, we show that $(m + 2)$ 1's in **M** cannot force more than three lengths.

Assume there are $(m + 2)$ 1's in **M**, $m - h$ of which are on the main diagonal. Since the desired result is obvious if $h = 0$, assume that $h > 0$, and let P_1 be a maximum-length strict path formed from off-diagonal 1's in the manner shown in Fig. 10.5.

Let $h_1 + 1$ be the number of off-diagonal 1's in P_1, so that h_1 0's on the main diagonal have their rows and columns taken care of by the 1's in P_1. If $h_1 = h$, then all but one of the off-diagonal 1's are in P_1, and it follows from the proof for $v(3, m) = m + 2$ that no more than three lengths are needed.

Assume henceforth that $h_1 < h$. Then there are $h_2 = h - h_1$ main-diagonal 0's not covered by P_1, and $h_2 + 1$ off-diagonal 1's not in P_1. These $h_2 + 1$ 1's must form a second strict path P_2 to take care of the h_2 main-diagonal 0's not covered by P_1. Only the initial 1 in P_2 can be in a row that has a 1 from P_1, and only the last 1 in P_2 can be in a column that has a 1 from P_1. In the first instance, the entry on the main diagonal to the left of the two 1's will be 1 if both P_1 and P_2 begin in that row and will be 0 otherwise (P_1 begins before P_2); in the second instance, the main-diagonal entry below the two 1's will be 1 if P_1 and P_2 end in that column, and 0 otherwise (P_1 ends after P_2).

Let A denote the set of off-diagonal 1's that are not dominated in the sense of $\subset_0$, and let B be the set of the other off-diagonal 1's, so $A \cup B$ consists of the 1's in P_1 and P_2. Each 1 in B is southwest of a 1 in A that dominates it: the dominating 1 must have a longer length in an interval representation. If $B = \emptyset$, then the off-diagonal 1's form a semiorder, which can be represented by one length according to Theorem 2.9. In this case **M** can be represented by two lengths. Moreover, if either P_i is included in B (all 1's in this P_i being dominated by a single 1 in A), then the proof of $v(3, m) = m + 2$ implies that **M** can be represented by three lengths.

Assume henceforth that each P_i has a 1 in A and that $B \neq \varnothing$. The 1's in A form a staircase pattern from upper left to lower right with a 1 at each outer corner. If P_2 begins in a P_1 row, then there are two 1's from A on that horizontal segment; if P_2 ends in a P_1 column, there are two 1's from A on that vertical segment. Otherwise, apart from main-diagonal 1's at the outer terminations of the staircase, each segment has precisely one 1 in A. In addition, the 1's in A going down the staircase alternate between P_1 and P_2 in the region where both are involved.

We now show that lengths N, 1, and $\delta > 0$ with N much larger than 1, and δ near 0, yield a valid representation of **M**.

Suppose first that no row or column has two 1's in A. For P_1, assign length N to each 1 in A, length 1 to each 1 in B, length 1 to the main-diagonal termini of P_1, and join the resultant intervals end-to-end. Do the same thing for P_2. Fix the interval sequence for P_1 on the line, then position the interval sequence for P_2 so that the right endpoint for a $1 \in A$ that is followed down the staircase by a $1 \in A$ from the other P_i is approximately in the middle of the latter's interval. This can be done for all such pairs of 1's in A since $N \gg 1$ and the 1's in A alternate between the P_i in the common region. Moreover, it guarantees that the length-1 intervals for B will be near the centers of the 1's in A that dominate them. It follows that all intervals for 1's in $A \cup B$, plus the four for the main-diagonal termini of the P_i are satisfactorily positioned.

None of the remaining $(m - h - 4)$ 1's on the main diagonal has another 1 in its row or column. We assign length δ to each of these and position their intervals in the obvious way to obtain a three-lengths representation for **M**.

Suppose henceforth that some row or column has two 1's in A. Assume for definiteness that this is true for row i, so that P_2 begins in row i. If P_2 also ends in a column used by P_1, then that part of their common region is treated symmetrically to the ensuing treatment of the start of their common region. If P_2 does not end in a P_1 column, the lower parts of the P_j are treated in essentially the same way as the case discussed in the preceding paragraphs.

With two 1's from A in row i, the first of these can be in either P_1 or P_2. The second is in the other P_j, and succeeding 1's from A alternate between the two strict paths until one of them ends. Beginning with the $1 \in A$ in row i for P_1, assign length N to the 1's in A and length 1 to the 1's in B, with intervals end-to-end as one goes down P_1. Do this also for P_2, then put the two together so that the right endpoint of each interval for a $1 \in A$ is about in the middle of its $1 \in A$ successor from the other P_j. It is easily checked that the off-diagonal 1's in row i and below are correctly positioned for a representation. (If P_2 does not end in a P_1 column, use length-1 intervals for their lower main-diagonal termini.)

Next, let

- $1°$ denote the 1 on the main diagonal at row i if P_1 begins in row i; otherwise,
- $1°$ is the $1 \in A$ for P_1 in column i above the 0 in cell (i, i).

Departures from Linearity

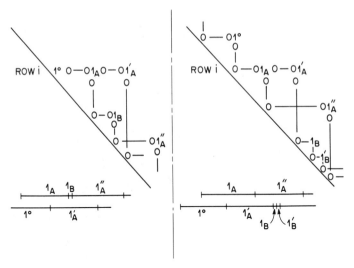

Figure 10.6

In either case, assign length N to $1°$ and position its right endpoint at the left endpoint of the interval for the *second* $1 \in A$ in row i. This is pictured in Fig. 10.6. In both cases, the interval for $1°$ intersects those for the two $1 \in A$ in row i, as it must for a representation, but it is wholly to the left of off-diagonal intervals in later rows. Moreover, in the second alternative for $1°$ (right part of Fig. 10.6), its interval begins to the left of the intervals for row i so that intermediate main-diagonal 1's can be properly positioned. In this case, intervals for the 1's in P_1 above $1°$ are constructed in the usual way.

Finally, main-diagonal 1's not involved at the termini of the P_j are assigned the small length δ and positioned among the others for a three-lengths representation of **M**. □

10.4 DEPARTURES FROM LINEARITY

The nice linear pattern for $v(k, m)$ described in Theorem 4(a) for $k \le 4$ breaks down at $k = 5$. The series $v(5, m)$ for $m = 9, 10, \ldots$ begins as $m + 4$, but after a while it changes to $m + 3$. In particular:

Theorem 5. $v(5, 29) = 32$.

Proof. Figure 10.7 shows the characteristic matrix (0's omitted) for a magnitude-29 interval order with 32 points. All off-diagonal 1's are covered by three strict paths (see Fig. 10.5), but the interval order cannot be represented by four lengths.

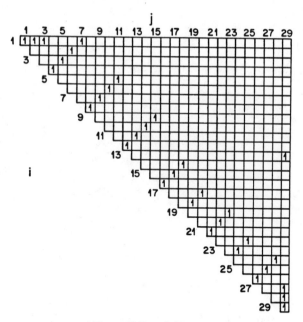

Figure 10.7 $v(5, 29) = 32$.

Suppose otherwise, and let $\rho(i, j)$ be the length of the interval for the 1 in cell (i, j). Then the 1's in the upper part of the matrix require

$$\rho(2, 6) + \rho(6, 10) + \rho(10, 14) > \rho(5, 11),$$

and the 1's in the lower part require

$$\rho(13, 29) > \rho(14, 18) + \rho(19, 23) + \rho(24, 28).$$

Let α be the longest of the four lengths, and β the next longest. The four-point $\subset_0$ chains ($\nearrow$) then imply that the two preceding inequalities are $3\beta > \alpha$ and $\alpha > 3\beta$.

Hence four lengths are not enough, so $v(5, 29) \leq 32$. Moreover, since $v(4, 29) = 32$ by Theorem 4(a), and $v(k, m)$ cannot decrease in k, $v(5, 29) = 32$. □

Before expanding on the theme of $v(k, m) < m + k - 1$, we review the preceding results for v. Some exact values of v for $m \leq 30$ are shown in Fig. 10.8. The first four rows come from Theorem 4(a), and the entries along the diagonal are from Theorem 4(b). The result of Theorem 5 is shown near the lower right, and it is easily seen that later entries in the fifth row have $v(5, m) = m + 3$.

An interesting aspect of $m = 29$ is that, because of $v(4, 29) = v(5, 29) = 32$, there must be a jump of at least 2 at some point as we go up column 29. This

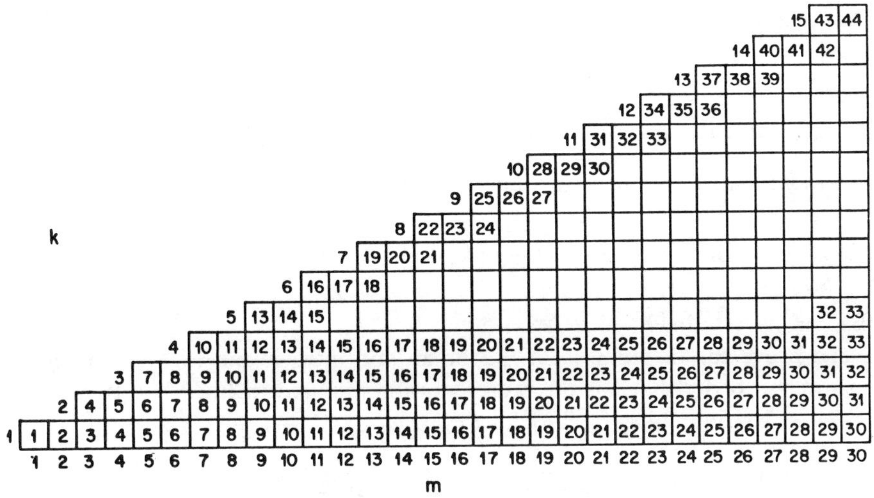

Departures from Linearity

Figure 10.8 $\nu(k, m)$.

could occur as early as $k = 6$ if $\nu(6, 29) = 34$. The numbers in the matrix, plus the obvious fact that $\nu(k, m)$ is nondecreasing in k, imply that $\nu(5, m) = \nu(5, m + 1)$ for some $m \leq 29$. Hence there are nontrivial m segments for some fixed k over which ν remains constant.

Additional information about ν and departures of $\nu(k, m)$ from $m + k - 1$ is provided by the limit function

$$\nu^*(k) = \lim_{m \to \infty} [\nu(k, m) - m].$$

Since a new column with a single 1 at the bottom can always be added to a magnitude-m characteristic matrix without changing the number of lengths needed for a representation, $\nu(k, m) - m$ is nonincreasing in m. Hence $\nu^*(k)$ is the smallest number of points needed beyond the basic m to force k lengths in representations of a magnitude-m interval order over all $m \geq 2k - 1$.

The following corollary is immediate from Theorems 4(a) and 5.

Corollary 1. $\nu^*(k) = k - 1$ for $k \leq 3$; $\nu^*(4) = \nu^*(5) = 3$.

In addition, Theorem 2 implies that

$$\nu^*(k) \leq k - 1 \quad \text{for all } k,$$

and Theorem 3 implies that

$$\nu^*(k) \leq \nu^*(k + 1) \leq \nu^*(k) + 1 \quad \text{for all } k.$$

Our final theorem establishes a tighter upper bound on ν^* that is exact at $k = 5$.

Theorem 6. *If k is divisible by 5, then*
$$\nu^*(k) \le \tfrac{4}{5}k - 1.$$

Corollary 2. $(m + k - 1) - \nu(k, m)$ *is unbounded on* K^*.

Proof. Since the theorem yields
$$\lim_{m}[(m + k - 1) - \nu(k, m)] \ge \frac{k}{5} \text{ when 5 divides } k,$$
the corollary follows.

To prove the theorem, let $n \ge 3$ be an odd integer, and let $(X, \prec)$ denote an interval order that has the representation pictured in Fig. 10.9. It is based on $2n + 1$ strict paths, P_1 through P_{2n+1}, has
$$18n + 18 = 10\left(\frac{3n + 1}{2}\right) + 6\left(\frac{n + 1}{2}\right) + 10$$
points, and has magnitude (min. number of left endpoints)
$$16n + 17 = 5(2n + 1) + 4\left(\frac{3n + 1}{2}\right) + 10.$$

The final 10 in each count is for the separate shortest intervals, for 10 main-diagonal 1's in **M**. The number of points in $(X, \prec)$ minus $m(X, \prec)$ is $2n + 1$.

We claim that at least $5(n + 1)/2$ lengths are needed to represent $(X, \prec)$. Then
$$\nu(5(n + 1)/2, 16n + 17) \le 18n + 18.$$

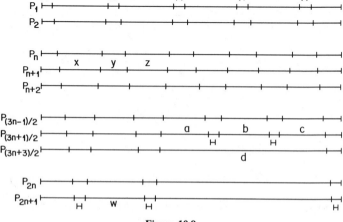

Figure 10.9

Hence $\nu^*(5(n+1)/2) \leq 2n+1$, or $\nu^*(k) \leq 4k/5 - 1$, for each k divisible by 5.

Suppose to the contrary of our claim that $5(n+1)/2 - 1 = (5n+3)/2$ lengths suffice to represent $(X, \prec)$. With no loss in generality, the 10 main-diagonal points can be given the shortest length. This leaves $(5n+1)/2$ longer lengths, say $\rho(1) < \rho(2) < \cdots < \rho((5n+1)/2)$.

Consider $\{x, y, z, w\}$ on the left of Fig. 10.9. Because w is the longest interval in a C_0 chain of $2n+1$ points (ignoring the main diagonal), $\rho(w) \geq \rho(2n+1)$. Moreover, because there must be $(2n+1) - (n+2) + 1 = n$ lengths longer than $\rho(y)$, and since $n + (3n+1)/2 = (5n+1)/2$, $\rho(y) \leq \rho((3n+1)/2)$. The same reasoning applies to x and z in their C_0 chains, so $\max\{\rho(x), \rho(z)\} \leq \rho((3n+1)/2)$. Since w is properly included in the connected span of x, y, and z,

$$\rho(2n+1) \leq \rho(w) < \rho(x) + \rho(y) + \rho(z) \leq 3\rho((3n+1)/2).$$

Hence $\rho(2n+1) < 3\rho((3n+1)/2)$.

Now consider $\{a, b, c, d\}$ on the right of the figure. Clearly, $\rho(d) > \rho(a) + \rho(b) + \rho(c)$. Since $(n-1)/2 = (2n+1) - (3n+3)/2$ lengths exceed $\rho(d)$, $\rho(d) \leq \rho(2n+1)$. Moreover, $\min\{\rho(a), \rho(b), \rho(c)\} \geq \rho((3n+1)/2)$, and therefore

$$3\rho\left(\frac{3n+1}{2}\right) < \rho(2n+1),$$

a contradiction to the conclusion of the preceding paragraph. □

10.5 MORE CONJECTURES

Since the theorems in this chapter provide only partial descriptions of σ and ν, many questions about these functions remain unanswered. Similar questions apply to $\tilde{\sigma}$ and $\tilde{\nu}$ for interval graphs in addition to whether $\tilde{\sigma} = \sigma$ and $\tilde{\nu} = \nu$. Theorem 7.2 for ρ-sets may help to settle these questions although it has not been used in the present chapter.

It is easily seen that Conjecture 1 is true if the following holds:

Conjecture 1'. *It is always possible to remove three points from an interval order $(X, \prec)$ with $|X| \geq 4$ in such a way that some minimum-lengths representation for the reduced interval order can be used to construct a representation for $(X, \prec)$ that requires at most one more length.*

The bound $2k \leq \sigma(k)$ for $k \geq 2$ reflects the fact that Conjecture 1' is true when "remove three points" is replaced by "remove two points." Short of verifying Conjecture 1, can the lower bound of $2k$ be increased significantly?

We have already noted that Conjecture 2 implies Conjecture 1. In fact, Conjecture 1 is equivalent to the weakening of Conjecture 2 which asserts that

$$\nu\left(\left\lfloor\frac{m+1}{2}\right\rfloor, m\right) \le \nu\left(\left\lfloor\frac{m+1}{2}\right\rfloor, m'\right) \quad \text{for all } m' > m.$$

Other weakenings of Conjecture 2 that imply Conjecture 1 include

C2$_1$. $\nu(k, m) \le \nu(k, m + 2)$
C2$_2$. $\nu(k, m) \le \nu(k, m + 3)$
C2$_3$. $\nu(k, m) + 2 \le \nu(k + 2, m + 1)$
C2$_4$. $\nu(k + 1, m) \le \nu(k, m) + 3$.

Verification of any of these would be a substantial move toward Conjecture 2. Note that C2$_4$ concerns the size of jumps as we go up a column in Fig. 10.8. It is not presently known whether there is a finite integer c such that $\nu(k + 1, m) \le \nu(k, m) + c$ for all $(k + 1, m) \in K^*$.

Additional unanswered questions include

1. What is the smallest m where $\nu(5, m) = m + 3$? (See Theorem 5.)
2. Is $\nu(k, m) = \nu(k, m + 1) = \nu(k, m + 2)$ for some $(k, m) \in K^*$?
3. Is $\nu(k, 2k + 2) = 3k + 1$ for all k? (See Theorem 4(b).)
4. Is there a substantially better upper bound on ν^* than the one in Theorem 6?
5. Is there a lower bound on $\nu^*(k)$ that becomes arbitrarily large as k increases?
6. What is the value of $\nu^*(k)$ for each $k \ge 6$?

References

Adams, E. W. (1965), Elements of a theory of inexact measurement, *Philos. Sci.* **32**, 205–228.

Altwegg, M. (1950), Zur Axiomatik der teilweise geordneten Menger, *Commen. Math. Helv.* **24**, 149–155.

Armstrong, W. E. (1939), The determinateness of the utility function, *Econ. J.* **49**, 453–467.

Armstrong, W. E. (1948), Uncertainty and the utility function, *Econ. J.* **58**, 1–10.

Armstrong, W. E. (1950), A note on the theory of consumer's behaviour, *Oxford Econ. Pap.* **2**, 119–122.

Baker, K. A. (1961), Dimension, join-independence, and breadth in partially ordered sets, unpublished manuscript.

Baker, K. A., Fishburn, P. C., and Roberts, F. S. (1970), Partial orders of dimension 2, interval orders and interval graphs, Paper P-4376, The Rand Corporation.

Baker, K. A., Fishburn, P. C., and Roberts, F. S. (1972), Partial orders of dimension 2, *Networks* **2**, 11–28.

Bayes, T. (1763), An essay towards solving a problem in the doctrine of chances, *Philos. Trans. R. Soc.* **53**, 370–418. Reprinted in *Facsimilies of Two Papers of Bayes* (ed. W. E. Deming), The Graduate School, Department of Agriculture, Washington, D.C., 1940.

Benzer, S. (1959), On the topology of the genetic fine structure, *Proc. Natl. Acad. Sci.* **45**, 1607–1620.

Berge, C. (1973), *Graphs and Hypergraphs*, North-Holland Publishing Co., Amsterdam.

Blumenthal, L. M. (1961), *A Modern View of Geometry*, Freeman, San Francisco.

Bogart, K. P. (1970), Decomposing partial orderings into chains, *J. Combinatorial Theory* **9**, 97–99.

Bogart, K. P. (1973a), Preference structures I: distances between transitive preference relations, *J. Math. Sociol.* **3**, 49–67.

Bogart, K. P. (1973b), Maximal dimensional partially ordered sets I. Hiraguchi's theorem, *Discrete Math.* **5**, 21–31.

Bogart, K. P., Rabinovitch, I., and Trotter, W. T., Jr. (1976), A bound on the dimension of interval orders, *J. Combinatorial Theory* **21**, 319–328.

Bogart, K. P., and Trotter, W. T., Jr. (1973), Maximal dimensional partially ordered sets II. Characterization of $2n$-element posets with dimension n, *Discrete Math.* **5**, 33–43.

Cantor, G. (1895), Beiträge zur Begründung der transfiniten Mengenlehre, *Math. Ann.* **46**, 481–512.

Cayley, A. (1859), On the analytical forms called trees, *Philos. Mag.* **18**, 374–378.

Chipman, J. S. (1971), Consumption theory without transitive indifference, in *Preferences, Utility, and Demand*, J. S. Chipman, L. Hurwicz, M. K. Richter, and H. F. Sonnenschein, Eds., Harcourt Brace Jovanovich, New York, pp. 224–253.

Chvátal, V., and Hammer, P. L. (1973), Set-packing and threshold graphs, University of Waterloo Research Report, CORR 73-21.

Cohen, J. E. (1978), *Food Webs and Niche Space*, Princeton University Press, Princeton, New Jersey.

Dean, R. A., and Keller, G. (1968), Natural partial orders, *Can. J. Math.* **20**, 535–554.

de Finetti, B. (1937), La prévision: ses lois logiques, ses sources subjectives, *Ann. l'Inst. Henri Poincaré* **7**, 1–68. Translated by H. E. Kyburg in Kyburg and Smokler (1964).

Dilworth, R. P. (1950), A decomposition for partially ordered sets, *Ann. Math.* **51**, 161–166.

Dirac, G. A. (1961), On rigid circuit graphs, *Abb. Math. Sem. Univ. Hamburg* **25**, 71–76.

Dushnik, B., and Miller, E. W. (1941), Partially ordered sets, *Am. J. Math.* **63**, 600–610.

Farkas, J. (1902), Theorie der einfachen Ungleichungen, *J. Reine Angew. Math.* **124**, 1–27.

Fishburn, P. C. (1969), Weak qualitative probability on finite sets, *Ann. Math. Stat.* **40**, 2118–2126.

Fishburn, P. C. (1970a), Intransitive indifference with unequal indifference intervals, *J. Math. Psychol.* **7**, 144–149.

Fishburn, P. C. (1970b), *Utility Theory for Decision Making*, Wiley, New York. Reprinted by Krieger, Huntington, New York, 1979.

Fishburn, P. C. (1970c), Utility theory with inexact preferences and degrees of preference, *Synthese* **21**, 204–221.

Fishburn, P. C. (1971), Betweenness, orders and interval graphs, *J. Pure Appl. Algebra* **1**, 159–178.

Fishburn, P. C. (1973a), Interval representations for interval orders and semiorders, *J. Math. Psychol.* **10**, 91–105.

Fishburn, P. C. (1973b), On the construction of weak orders from fragmentary information, *Psychometrika* **38**, 459–472.

Fishburn, P. C. (1978), Operations on binary relations, *Discrete Math.* **21**, 7–22.

Fishburn, P. C. (1981a), Restricted thresholds for interval orders: a case of nonaxiomatizability by a universal sentence, *J. Math. Psychol.* **24**, 276–283.

Fishburn, P. C. (1981b), Maximum semiorders in interval orders, *SIAM J. Algebraic Discrete Methods* **2**, 127–135.

Fishburn, P. C. (1982), Aspects of semiorders within intervals orders, *Discrete Math.* **40**, 181–191.

Fishburn, P. C. (1983a), Threshold-bounded interval orders and a theory of picycles, *SIAM J. Algebraic Discrete Methods* **4**, 290–305.

Fishburn, P. C. (1983b), Numbers of lengths for representations of interval orders, in *Progress in Combinatorial Optimization*, W. R. Pulleyblank, Ed., Academic Press (1984), New York.

Fishburn, P. C. (1983c), Interval lengths for interval orders: a minimization problem, *Discrete Math.* **47**, 63–82.

Fishburn, P. C. (1984a), A characterization of uniquely representable interval graphs, *Discrete Appl. Math.* (to appear).

Fishburn, P. C. (1984b), Interval graphs and interval orders, *Discrete Math.* (to appear).

Fishburn, P. C. (1984c), Paradoxes of two-length interval orders, *Discrete Math.* (to appear).

Fishburn, P. C., and Gehrlein, W. V. (1974), Alternative methods of constructing strict weak orders from interval orders, *Psychometrika* **39**, 501–516.

Fishburn, P. C., and Gehrlein, W. V. (1975), A comparative analysis of methods for constructing weak orders from partial orders, *J. Math. Sociol.* **4**, 93–102.

Fulkerson, D. R., and Gross, O. A. (1965), Incidence matrices and interval graphs, *Pac. J. Math.* **15**, 835–855.

Ghouilà-Houri, A. (1962), Caractérisation des graphes non orientes dont on peut orienter les arêtes de manière à obtenir le graphe d'une relation d'ordre, *C. R. Acad. Sci. Paris* **254**, 1370–1371.

References

Gilmore, P. C., and Hoffman, A. J. (1962), Characterizations of comparability and interval graphs (abstract), *International Congress of Mathematicians*, Stockholm, p. 29.

Gilmore, P. C., and Hoffman, A. J. (1964), A characterization of comparability graphs and of interval graphs, *Can. J. Math.* **16**, 539–548.

Gleason, A. M., and Greenwood, R. E. (1955), Combinatorial relations and chromatic graphs, *Can. J. Math.* **7**, 1–7.

Golumbic, M. C. (1980), *Algorithmic Graph Theory and Perfect Graphs*, Academic Press, New York.

Goodman, N. (1951), *The Structure of Appearance*, Harvard University Press, Cambridge, Massachusetts.

Graham, R. L., Rothschild, B. L., and Spencer, J. H. (1980), *Ramsey Theory*, Wiley, New York.

Gross, O. A. (1962), Preferential arrangements, *Am. Math. Mon.* **69**, 4–8.

Hajös, G. (1957), Über eine Art von Graphen, *Int. Math. Nachr.* **11**, 65.

Hanlon, P. (1982), Counting interval graphs, *Trans. Am. Math. Soc.* **272**, 383–426.

Harary, F. (1964), A graph theoretic approach to similarity relations, *Psychometrika* **29**, 143–151.

Harzheim, E. (1970), Ein Endlichkeitssatz über die Dimension teilweise geordneter Mengen, *Math. Nachr.* **46**, 183–188.

Hilbert, D. (1899), *Grundlagen der Geometrie*, Teubner, Leipzig.

Hiraguchi, T. (1955), On the dimension of orders, *Sci. Rep. Kanazawa Univ.* **4**, 1–20.

Huntington, E. V. (1924), A new set of postulates for betweenness, with proof of complete independence, *Trans. Am. Math. Soc.* **26**, 257–282.

Huntington, E. V., and Kline, J. R. (1917), Sets of independent postulates for betweenness, *Trans. Am. Math. Soc.* **18**, 301–325.

Kelley, J. L. (1955), *General Topology*, American Book Company, New York.

Kelly, D., and Rival, I. (1975), Certain partially ordered sets of dimension three, *J. Combinatorial Theory* (A) **18**, 239–242.

Kemeny, J. G. (1959), Mathematics without numbers, *Daedalus* **88**, 577–591.

Kemeny, J. G., and Snell, J. L. (1962), *Mathematical Models in the Social Sciences*, Ginn, Boston, pp. 11–19.

Kendall, D. G. (1969), Incidence matrices, interval graphs and seriation in archaeology, *Pac. J. Math.* **28**, 565–570.

Keynes, J. M. (1921), *A Treatise on Probability*, Macmillan, New York; second edition, 1929; Torchbook Edition, Harper and Row, 1962.

Klee, V. (1955), Separation properties of convex cones, *Proc. Am. Math. Soc.* **6**, 313–318.

Komm, H. (1948), On the dimension of partially ordered sets, *Am. J. Math.* **70**, 507–520.

Kraft, C. H., Pratt, J. W., and Seidenberg, A. (1959), Intuitive probability on finite sets, *Ann. Math. Stat.* **30**, 408–419.

Krantz, D. H., Luce, R. D., Suppes, P., and Tversky, A. (1971), *Foundations of Measurement*, Vol. 1, Academic Press, New York.

Kuhn, H. W. (1956), Solvability and consistency for linear equations and inequalities, *Am. Math. Mon.* **63**, 217–232.

Kuhn, H. W., and Tucker, A. W. (Eds.) (1956), *Linear Inequalities and Related Systems*, Annals of Mathematics Study 38, Princeton University Press, Princeton, New Jersey.

Kyburg, H. E., Jr., and Smokler, H. E. (Eds.) (1964), *Studies in Subjective Probability*, Wiley, New York.

Laplace, Pierre Simon de (1812), *Théorie Analitique des Probabilités*, Paris. Reprinted in *Oeuvres Complètes*, Vol. 7 (Paris, 1974). See also *Essai Philosophique sur les Probabilités* (Paris, 1814), translated as *A Philosophical Essay on Probabilities* (Dover, New York, 1951).

Leibowitz, R. (1978), Interval counts and threshold numbers of graphs, Ph.D. Dissertation, Rutgers University.

Lekkerkerker, C. G., and Boland, J. Ch. (1962), Representation of a finite graph by a set of intervals on the real line, *Fundam. Math.* **51**, 45–64.

Luce, R. D. (1956), Semiorders and a theory of utility discrimination, *Econometrica* **24**, 178–191.

Luce, R. D., and Suppes, P. (1965), Preference, utility, and subjective probability, in *Handbook of Mathematical Psychology*, III, R. D. Luce, R. R. Bush, and E. Galanter, Eds., Wiley, New York, pp. 249–410.

Mendelson, E. (1982), Races with ties, *Math. Mag.* **55**, 170–175.

Mirsky, L., and Perfect, H. (1966), Systems of representatives, *J. Math. Anal. Appl.* **15**, 520–568.

Moon, J. W. (1968), *Topics on Tournaments*, Holt, Rinehart and Winston, New York.

Pasch, M. (1882), *Vorlesungen über neuere Geometrie*, Teubner, Leipzig.

Peano, G. (1889), *I Principii di Geometria*, Turin.

Peano, G. (1894), Sui fondamenti della geometria, *Riv. Mat.* **4**, 51–90.

Perles, M. A. (1963a), A proof of Dilworth's decomposition theorem for partially ordered sets, *Isr. J. Math.* **1**, 105–107.

Perles, M. A. (1963b), On Dilworth's theorem in the infinite case, *Isr. J. Math.* **1**, 108–109.

Rabinovitch, I. (1977), The Scott-Suppes theorem on semiorders, *J. Math. Psychol.* **15**, 209–212.

Rabinovitch, I. (1978a), The dimension of semiorders, *J. Combinatorial Theory* (A) **25**, 50–61.

Rabinovitch, I. (1978b), An upper bound on the "Dimension of interval orders," *J. Combinatorial Theory* (A) **25**, 68–71.

Rado, R. (1949), Axiomatic treatment of rank in infinite sets, *Can. J. Math.* **1**, 337–343.

Ramsey, F. P. (1931), Truth and probability, in *The Foundations of Mathematics and Other Logical Essays*, Harcourt, Brace, New York. Reprinted in Kyburg and Smokler (1964), 61–92.

Reid, K. B., and Parker, E. T. (1970), Disproof of a conjecture of Erdös and Moser on tournaments, *J. Combinatorial Theory* **9**, 225–238.

Roberts, F. S. (1969), Indifference graphs, in *Proof Techniques in Graph Theory*, F. Harary, Ed., Academic Press, New York, pp. 139–146.

Roberts, F. S. (1970), On nontransitive indifference, *J. Math. Psychol.* **7**, 243–258.

Roberts, F. S. (1971), On the compatibility between a graph and a simple order, *J. Combinatorial Theory* **11**, 28–38.

Roberts, F. S. (1979), *Measurement Theory*, Addison-Wesley, Reading, Massachusetts.

Robinson, G. de B. (1940), *The Foundations of Geometry*, University of Toronto Press, Toronto.

Rockafellar, R. T. (1970), *Convex Analysis*, Princeton University Press, Princeton, New Jersey.

Rosenstein, J. G. (1982), *Linear Orderings*, Academic Press, New York.

Savage, L. J. (1954), *The Foundations of Statistics*, Wiley, New York. Second revised edition, Dover, New York, 1972.

Scott, D. (1964). Measurement structures and linear inequalities, *J. Math. Psychol.* **1**, 233–247.

Scott, D., and Suppes, P. (1958), Foundational aspects of theories of measurement, *J. Symb. Logic* **23**, 113–128.

Sholander, M. (1952), Trees, lattices, order and betweenness, *Proc. Am. Math. Soc.* **3**, 369–381.

Shuchat, A. (1984), Matrix and network models in archaeology, *Math Mag.* **57**, 3–14.

Świstak, P. (1980), Some representation problems for semiorders, *J. Math. Psychol.* **21**, 124–135.

Szpilrajn, E. (1930), Sur l'extension de l'ordre partiel, *Fundam. Math.* **16**, 386–389.

Titiev, R. J. (1972), Measurement structures in classes that are not universally axiomatizable, *J. Math. Psychol.* **9**, 200–205.

Trotter, W. T., Jr. (1974), Irreducible posets with large height exist, *J. Combinatorial Theory* **17**, 337–344.

References

Trotter, W. T., Jr. (1975), Inequalities in dimension theory for posets, *Proc. Am. Math. Soc.* **47**, 311–316.

Tverberg, H. (1967), On Dilworth's decomposition theorem for partially ordered sets, *J. Combinatorial Theory* **3**, 305–306.

Veblen, O. (1904), A system of axioms for geometry, *Trans. Am. Math. Soc.* **5**, 343–384.

Wiener, N. (1914), A contribution to the theory of relative position, *Proc. Camb. Philos. Soc.* **17**, 441–449.

Wine, R. L., and Freund, J. E. (1957), On the enumeration of decision patterns involving n means, *Ann. Math. Stat.* **28**, 256–259.

Wolk, E. S. (1965), A note on "The comparability graph of a tree," *Proc. Am. Math. Soc.* **16**, 17–20.

Index

Adams, E. W., 130, 205
Adjacency, 9
Altwegg, M., 59, 205
Antichain, 12
   maximal, 12
   maximum, 12
   restricted, 116
Armstrong, W. E., 20, 205
Asteroidal triple, 36, 46
Astral triple, 51
Asymmetry, 1, 2
Axiom A $[p, q]_n$, 150
Axiom of choice, 7

Baker, K. A., 80, 84–87, 95, 96, 205
Bayes, T., 130, 205
Benzer, S., 37, 205
Berge, C., 9, 15, 205
Betweenness relation, 57
   agreement with, 58, 70, 74
   complete, 59
   graphs induced by, 58
   strict, 58
   string in, 61
   strong, 58
   weak pair in, 62
Binary relation, 1
   asymmetric part of, 2
   complement of, 2
   composition of, 1
   dual of, 2
   operations on, 2
   properties of, 1
   symmetric part of, 2
Bipartite graph, 14, 38
Blumenthal, L. M., 59, 205
Bogart, K. P., 13, 33, 79–81, 84, 92, 94, 95, 205
Boland, J. Ch., 36, 46, 47, 49, 51, 208
Breadth:
   of interval order, 96
   of poset, 77

Buried subgraph, 54
   K-set of, 54
Bush, R. R., 208

Cantor, G., 17, 133, 205
Cantor's theorem, 133
Cayley, A., 98, 205
Chain, 3
   maximal, 3
   maximum, 3
Characteristic matrix, 26, 186
   strict path in, 196
Chipman, J. S., 18, 20, 205
Choice function, 9
Chord, 14
Chvátal, V., 173, 205
Clique, 14
   end, 48
Clique ordering, 41
   consecutive, 45
   proper consecutive, 50
Cohen, J. E., 37, 206
Comparability graph, 15
Complete graph, 13, 46
Completeness, 1
Complex open gap, 42
Component, 14
Composition, 1, 21, 149
Conjectures on lengths, 187, 203, 204
Conjoint weak order, 25, 134
Connected points, 14
Consecutive-1's property, 40
Convex cone, 124
   open, 129
Convex hull, 124
Crown, 87
Cycle, 9, 14
   chord of, 14
   simple, 14
   triangular chord of, 14
Cyclic order, 59

Dean, R. A., 98, 206
Deficient point, 42
de Finetti, B., 130, 206
Deming, W. E., 205
Dense relation, 21
Dense subset, 133
Depth:
  of interval order, 96
  of poset, 77, 125
Diagram, 21
Dilworth, R. P., 12, 80, 83, 206
Dilworth's theorem, 12, 13, 84
Dimension:
  of interval order, 92, 95
  of lattice, 79
  of poset, 8, 76
  of semiorder, 89
Dirac, G. A., 47, 206
Dushnik, B., 79, 85, 206

Empty relation, 2
Equivalence class, 1
Equivalence relation, 1
Event, 130

Fan, Ky, 124
Farkas, J., 124, 206
Fence, 87
Finite-lengths classes, 170
Fishburn, P. C., 2, 33, 59, 85–87, 95, 96, 130, 133, 205, 206
Food webs, 37
Freund, J. E., 98, 209
Fulkerson, D. R., 37, 40, 45, 206

Galanter, E., 208
Gehrlein, W. V., 33, 206
Ghouilà-Houri, A., 9, 15, 206
Gilmore, P. C., 15, 45, 46, 51, 207
Gleason, A. M., 101, 207
Global score function, 33
Goldman, A. J., 124
Golumbic, M. C., 37, 47, 173, 207
Goodman, N., 59, 207
Graham, R. L., 95, 102, 207
Graph, 13
  bipartite, 14
  comparability, 15
  complement of, 36
  complete, 13
  component, of, 14
  connected, 14
  incomparability, 85
  indifference, 35

  induced by betweenness, 58
  intersection, 37
  interval, 14, 35
  magnitude of, 77
  nonuniversally connected, 53
  properties of, 13
  simplicial point of, 47
  triangulated, 46
Greenwood, R. E., 101, 207
Gross, O. A., 37, 40, 45, 98, 206, 207

Hajös, G., 207
Hammer, P. L., 173, 205
Hanlon, P., 16, 54, 100, 145, 207
Harary, F., 37, 207
Harzheim, E., 87, 207
Hasse diagram, 3, 21
Height of poset, 76
Hilbert, D., 59, 207
Hiraguchi, T., 80, 81, 207
Hiraguchi's theorem, 79, 84
Hoffman, A. J., 15, 45, 46, 51, 207
Huntington, E. V., 16, 59, 207
Hurwicz, L., 205

Incomparability graph, 85
Incomparability relation, 85
Independent set, 14
Indifference, 20, 37
Indifference graph, 35, 37
  agreement with, 50
  maximum in interval graph, 102
  uniform representation of, 37
  uniquely orderable, 53
Intersection graph, 37
Interval, 12
  closed, 12
  free, 181
  open, 12
Interval graph, 14, 35
  agreement with, 39
  for betweenness relation, 63
  buried subgraph of, 54
  characterizations of, 46
  length-bounded, 144, 152
  magnitude of, 39
  number of, 100
  number of lengths in, 170, 188
  representation of, 36
  uniquely orderable, 53
  unit, 100
Interval order, 18
  agreeing, 39
  betweenness agreeing, 62

Index

breadth of, 96
characteristic matrix of, 26, 186
closed real representation of, 135
conjoint weak order of, 25
depth of, 96
depth-2, 176
dimension of, 92, 95
hierarchical series of, 114
left-justified, 130
length-bounded, 144, 150
magnitude of, 26, 186
monotonic, 130
number of, 100
number of lengths, in, 170, 187
real representation of, 134, 140
representation of, 30
strong, 86
two-lengths representations of, 179
Irreflexivity, 1
Isolated point, 14
Isomorphism, 3

Join-independent subset, 77
Just-noticeable-difference, 20

Keller, G., 98, 206
Kelley, J. L., 7, 207
Kelly, D., 88, 207
Kemeny, J. G., 33, 207
Kendall, D. G., 37, 207
Keynes, J. M., 20, 207
Klee, V., 124, 207
Kline, J. R., 59, 207
Komm, H., 79, 207
Kraft, C. H., 130, 207
Krantz, D. H., 130, 133, 207
K-set, 54
Kuhn, H. W., 124, 207
Kuratowski's lemma, 7
Kyburg, H. E., 206, 207

Laplace, P. S. de, 130, 207
Lattice of subsets, 32, 79
Leibowitz, R., 173, 208
Lekkerkerker, C. G., 36, 46, 47, 49, 51, 208
Length-bounded interval graph, 144, 152
Length-bounded interval order, 144, 150
Length function, 30
 strictly positive, 126, 144
Lengths, numbers of, 170, 187
Lexicographic extension, 88
Linear extension, 7
 as lower extension, 80
 nonseparating, 85
 realizing, 80
 as upper extension, 81
Linear inequalities, 124
Linearly ordered subset, 3
Linear order, 3
 betweenness agreeing, 74
 of maximal cliques, 40
 order dense subset of, 133
Linear solution theory, 123
Location function, 30, 126
Loop, 13
Luce, R. D., 20, 207, 208

Magnitude, 26, 39, 77, 186
Mendelson, E., 98, 208
Miller, E. W., 79, 85, 206
Mirski, L., 9, 208
Moon, J. W., 101, 208

Negative transitivity, 1, 2
Nonuniversally connected graph, 53

Order, 3
 clique, 41
 cyclic, 59
 interval, 18
 linear, 3
 partial, 3
 semi-, 18
 semitransitive, 18
 weak, 3
Order denseness, 133

Parker, E. T., 101, 128
Partial order, 3
 breadth of, 77
 crown, 87
 dense, 21
 depth of, 77, 125
 dimension of, 8, 76
 fence, 87
 height of, 76
 irreducible, 88
 join-independent subset of, 77
 magnitude of, 77
 sequel of, 4, 31
 weak order extension of, 30
 width of, 76
Partial semiorder, 18
Pasch, M., 59, 208
Path, 14
 simple, 14
 strict, 196
Peano, G., 59, 208

Perfect, H., 9, 208
Perles, M. A., 12, 84, 208
Petrie, F., 41
Petrie matrix, 41
Picycle, 153
  excesses of, 153
  forbidden, 154
  index of, 153
  length of, 158
  reducible, 158
  simple, 154
  transition in, 156
Poset, 3. *See also* Partial order
Pratt, J. W., 130, 207
Precedence, 19
Preference, 20
Probability, 20, 130
Probability measure, 131
Psychophysical measurement, 20
Pulleyblank, W. R., 206

Rabinovitch, I., 86, 88–90, 92, 94, 95, 122, 205, 208
Rado, R., 9, 208
Rado's theorem, 9, 11, 12
Ramsey, F. P., 130, 208
Ramsey theory, 95, 101
Redundant point, 42
Reflexivity, 1
Reid, K. B., 101, 208
Related set, 2
  number of, 97
  reorientation of, 8
Relation, 1
  binary, 1
  composition of, 1
  equivalence, 1
  incomparability, 85
  restriction of, 3
  similarity, 37
  ternary, 1, 57
Reorientation, 8
Representation:
  of indifference graph, 37
  of interval graph, 36
  of interval order, 28, 30
  length-restricted, 144, 179, 187
  monotonic, 142
  real, 122
  of semiorder, 29
Restriction operation, 3
Richter, M. K., 205
Rival, I., 88, 207

Roberts, F. S., 37, 38, 51, 53, 59, 85–87, 95, 96, 122, 205, 208
Robinson, G. de B., 59, 208
Rockafellar, R. T., 124, 208
Rosenstein, J. G., 90, 208
Rothschild, B. L., 95, 102, 207

Savage, L. J., 130, 208
Scott, D., 16, 30, 122, 129, 151, 208
Scott-Suppes theorem, 30, 122, 129
Seidenberg, A., 130, 207
Semiorder, 18
  agreeing, 50
  betweenness agreeing, 70
  closed real representation of, 136
  dimension of, 89
  maximum in interval order, 102
  number of, 98
  partial, 18
  real representation of, 140
  representation of, 29
  Scott-Suppes theorem for, 30
  uniform representation of, 37
Semitransitivity, 18
Separating hyperplane, 124
Separator, 47
Sequel of poset, 4, 31
  degree of, 31
Sholander, M., 59, 208
Shuchat, A., 41, 208
Simplicial point, 47, 135
  strong, 47
  weak, 47
Smokler, H. E., 207
Snell, J. L., 33, 207
Sonnenschein, H. F., 205
Spencer, J. H., 95, 102, 207
States, 130
String, 61
Subgraph (induced), 14
  buried, 54
Suppes, P., 16, 20, 30, 122, 151, 207, 208
Świstak, P., 140, 208
Symmetric complement, 2
Symmetry, 1, 2, 57
Szpilrajn, E., 7, 208
Szpilrajn's theorem, 7, 8, 77, 80, 89

Ternary relation, 1, 57
  symmetric, 57
Theorem of the alternative, 124
Titiev, R. J., 151, 208

Index

Tournament, 101
Transitive closure, 2
Transitive reorientation, 9
Transitivity, 1
  negative, 1
  pseudo-, 9
Triangular chord, 14
Triangulated graph, 46, 51
Trotter, W. T., Jr., 79, 80, 83, 84, 88, 92, 94, 95, 205, 208, 209
Tucker, A. W., 124, 207
Tverberg, H., 12, 209
Tversky, A., 207

Uniform representation, 37
Unitary class, 147
Unit interval graph, 100. *See also* Indifference graph
Universal point, 14

Universal relation, 2

Veblen, O., 59, 209
Vector, 123
  integral, 123
  rational, 123

Weak order, 3
  betweenness agreeing, 74
  conjoint, 25
  distance between, 33
  extension to, 30
  lexicographic extension of, 88
  number of, 98
  real representation of, 133
Width of poset, 76
Wiener, N., 19, 20, 209
Wine, R. L., 98, 209
Wolk, E. S., 9, 209